Multiple Criteria Decision Making and Risk Analysis Using Microcomputers

Edited by

Birsen Karpak

Youngstown State University, Department of Management
Youngstown, OH 44 555, USA

Stanley Zionts

School of Management
State University of New York at Buffalo
Buffalo, NY 14 260, USA

Springer-Verlag Berlin Heidelberg New York
London Paris Tokyo Hong Kong
Published in cooperation with NATO Scientific Affairs Division

Proceedings of the NATO Advanced Study Institute on Multiple Criteria Decision Making and Risk Analysis Using Microcomputers, held at Tarabya, Istanbul, Turkey, June 28–July 8, 1987.

ISBN-13: 978-3-642-74921-6 e-ISBN-13: 978-3-642-74919-3
DOI: 10.1007/978-3-642-74919-3

Library of Congress Cataloging-in-Publication Data. NATO Advanced Study Institute on Multiple Criteria Decision Making and Risk Analysis Using Microcomputers (1987 : Istanbul, Turkey) Multiple criteria decision making and risk analysis using microcomputers / edited by Birsen Karpak, Stanley Zionts. p. cm. — (NATO ASI series. Series F, Computer and systems sciences ; vol. 56) "Proceedings of the NATO Advanced Study Institute on Multiple Criteria Decision Making and Risk Analysis Using Microcomputers, held in Istanbul, Turkey June 28–July 8, 1987" — T.p. verso. "Published in cooperation with NATO Scientific Affairs Division."
ISBN-13: 978-3-642-74921-6 (U.S.)
1. Decision-making — Data processing — Congresses. 2. Risk assessment — Data processing — Congresses. 3. Microcomputers — Congresses. I. Karpak, Birsen. II. Zionts, Stanley, 1937– III. North Atlantic Treaty Organization. Scientific Affairs Division. IV. Title. V. Series: NATO ASI series. Series F, Computer and systems sciences ; vol. 56. T57.95.N376 1987 658.4'03 — dc20 89-19671

Softcover reprint of the hardcover 1st edition 1989

2145/3140-543210 – Printed on acid-free-paper

CONTENTS

Multiple Criteria Decision Making and Risk Analysis Using Microcomputers

NATO ASI Series

Advanced Science Institutes Series

A series presenting the results of activities sponsored by the NATO Science Committee, which aims at the dissemination of advanced scientific and technological knowledge, with a view to strengthening links between scientific communities.

The Series is published by an international board of publishers in conjunction with the NATO Scientific Affairs Division

A Life Sciences B Physics	Plenum Publishing Corporation London and New York
C Mathematical and Physical Sciences D Behavioural and Social Sciences E Applied Sciences	Kluwer Academic Publishers Dordrecht, Boston and London
F Computer and Systems Sciences G Ecological Sciences H Cell Biology	Springer-Verlag Berlin Heidelberg New York London Paris Tokyo Hong Kong

Series F: Computer and Systems Sciences Vol. 56

INTRODUCTION

Roughly nine years ago, the two editors met for the first time in Amsterdam, the Netherlands at the EURO III meeting (organized by the Association of European Operational Research Societies) there. As a result of our initial meeting, the two of us planned and carried out a number of activities in the multiple criteria decision making area, much of it supported by the North Atlantic Treaty Organization (NATO). The latest of these activities was a NATO Advanced Study Institute (ASI) on multiple criteria decision making and risk analysis using microcomputers. The institute was held in Tarabya, Istanbul, TURKEY, on June 28 - July 8, 1987.

We received over 100 applications from professors (and a few graduate students) in 13 countries. Roughly half of them were able to participate. The ASI was a great success! Substantial knowledge transfer and learning took place. In addition to the planned presentations, we had several panels and round table discussions. Though we had planned these in advance, we implemented them to fit the occasion, and also organized a few special sessions on site to respond to participants' interests.

Both of the editors have had a long - standing interest in multiple criteria models as well as computers. Organizing such a meeting was particularly appropriate because of the recent explosive growth in the development of both multicriteria models

NATO ASI Series, Vol. F56
Multiple Criteria Decision Making and Risk Analysis
Using Microcomputers. Edited by B. Karpak and S. Zionts

and microcomputers. Multiple criteria models have evolved over a period of roughly twenty years. Though computers have been widely used for many years, the technology of computers has been substantially enhanced. Because of the explosive growth in microcomputers, the microcomputer is now found almost everywhere. The dependence of multiple criteria decision models upon microcomputers has increased. Multicriteria models rely on tradeoffs, comparisons and extensive calculations. The tradeoffs can be effectively communicated by features of the microcomputer and the calculations can be quickly carried out. The result is an efficient marriage of methodology and technology.

We have organized the papers in this volume in a logical way. First is the overview by Stanley Zionts which reviews a number of methods of multiple criteria decision making which are of both historical and current interest. Next is a paper by James Dyer which provides an overview of some concepts of risk analysis. Specifically, if the utility function and value function are not the same, the relationship between them gives an insight into the effect of risk on preferences. Tversky and Kahneman in "Rational Choice and the Framing of Decisions" argue that most decision makers in practice violate certain principles of normative decision theory. Tversky and Kahneman also explore how important the proper framing of decisions is. Next is a paper by Shi and Yu that emphasizes the importance of understanding the human behavior mechanism to successfully solve nontrivial decision problems. People normally do not depart from their habitual ways of thinking. In a second paper the authors explore the concepts of satisficing and compromise solutions. They then explore ways in which an ideal solution derived based on these concepts can be used in decision making. In TRIMAP, Joao Climaco presents a clever method and microcomputer program for exploring nondominated solutions to three-criteria linear programming problems. He effectively uses computer graphics in doing this. In the next paper Ralph Steuer presents a method that uses the Tchebycheff norm for solving multiple objective linear programming problems. His approach has evolved from his earlier work in the field. Korhonen and Wallenius present VIG, a method for visual interactive goal programming. Their approach effectively interfaces computer graphics in a user friendly way, resulting in an impressive piece of software. In the next paper Koksalan and Beseli develop a variation of the Zionts - Wallenius approach to multiple objective linear programming. They use ideas suggested by Green and Srinivasan in doing this. In "Prioritization Techniques Based on Stochastic Paired Comparisons" V.R.R. Uppuluri explores the use of Saaty's Analytic Hierarchy Process to analyze data based on stochastic paired comparisons. The next two papers by Gunter Fandel and Holger Hegemann and by Birsen Karpak present applications of MCDM. Fandel and Hegemann describe an MCDM application in hospital planning in the Federal Republic of Germany. Considering the flows of patients through the hospital as flows in a network, the authors use queuing theory to facilitate hospital capacity planning. Karpak compares the contracting cone method, a modified two person zero-sum approach and Steuer's interactive

Tchebycheff method in an application to a production planning problem of a Turkish manufacturing company. In the last article, Aydin Ulkucu summarizes a panel discussion on conflicting objectives in software system design. In addition to the summary, he also includes the dialogue of participants.

We are pleased with the ASI and the results thereby achieved. It is difficult to organize an ASI involving many people in a number of different countries. With the help of many individuals and organizations we were able to do it. We would like to thank those who have helped us: the NATO Scientific Affairs Division, in particular Dr. Bulent Bayraktar and Dr. Craig Sinclair, for sponsoring the institute. We also want to thank Istanbul University, particularly Dr. Cem'i Demiroglu and Dr. Kemal Tosun, and the National Science Foundations of Portugal (INVOTAN), TURKEY (TUBITAK) and the USA for their support. The School of Management, State University of New York at Buffalo, and Department of Management at Youngstown State University have contributed to this effort in many ways including the time spent by the organizers as well as the contributions of secretaries, Mrs. Valerie Limpert and Mrs. Ruth Sing at Buffalo and Ms. Janet Mitchell at Youngstown. We also appreciate the support of our administrators, particularly Dean Joseph A. Alutto and Professor Clement Psenicka at Youngstown State.

We appreciate the help and support of the organizing committee, Paul Slovic of Decision Research, Peter Fishburn of AT&T Bell Laboratories, Jyrki Wallenius of University of Jyvaskyla and we acknowledge their contribution. Last but not least we wish to thank Ali Coskun, president of the union of chambers of commerce and industry of Turkey, for arranging participation of Turkish businessmen.

Birsen Karpak
Department of Management
Youngstown State University
Youngstown, OH 44555

Stanley Zionts
School of Management
State University of New York
Buffalo, NY 14260

February, 1989

MULTIPLE CRITERIA MATHEMATICAL PROGRAMMING:

AN UPDATED OVERVIEW AND SEVERAL APPROACHES

Stanley Zionts
Alumni Professor of Decision Support Systems
State University of New York at Buffalo
Buffalo, New York 14260 USA

Introduction

Multiple Criteria Decision Making (MCDM) refers to making decisions in the presence of multiple, usually conflicting, objectives. Multiple criteria decision problems pervade all that we do and include such public policy tasks as determining a country's policy developing a national energy plan, as well as planning national defense expenditures, in addition to such private enterprise tasks as new product development, pricing decisions, and research project selection. For an individual, the purchase of an automobile or a home exemplifies a multiple criteria problem. Even such routine decisions as the choice of a lunch from a menu, or the assignment of job crews to jobs constitute multiple criteria problems. All have a common thread--multiple conflicting objectives.

In this study, we discuss some of the important aspects of solving such problems, and present some methods developed for solving multiple criteria mathematical programming problems and discrete alternative models. We also discuss some applications of the methods.

In multiple criteria decision making there is a decision maker (or makers) who makes the decision, a set of objectives

NATO ASI Series, Vol. F56
Multiple Criteria Decision Making and Risk Analysis
Using Microcomputers. Edited by B. Karpak and S. Zionts

Constraint. A constraint is a limit on attributes and decision variables that may or may not be stated mathematically. For example, that a plant can be operated at most twelve hours per day is a constraint.

Structuring an MCDM Situation

Most problems have, in addition to multiple conflicting objectives a hierarchy of objectives. For example, according to Manheim and Hall (1967), the objective for evaluating passenger transportation facilities serving the Northeast Corridor of the U.S. in 1980 was "The Good Life." This superobjective was subdivided into four main objectives:

1. Convenience
2. Safety
3. Aesthetics
4. Economic Considerations

These in turn are divided into subobjectives, and so on forming a hierarchy of objectives.

Some of the objectives, such as economic considerations, have attributes that permit a precise performance measurement. Other, such as aesthetics, are highly subjective. Not wanting to convert the word subjective to a noun, we may, therefore, have a subjective objective. Further, the number of objectives may be large in total.

To adequately represent the objectives, we must choose appropriate attributes. Keeney and Raiffa (1976) indicate five characteristics the selected attributes of the objectives should have:

that are to be pursued, and a set of alternatives from which one is to be selected.

GOALS, CRITERIA, OBJECTIVES, ATTRIBUTES, CONSTRAINTS, AND TARGETS: THEIR RELATIONSHIPS

In a decision situation we have goals, criteria, objectives, attributes, constraints, and targets, in addition to decision variables. Although goals, criteria, and targets have essentially the same dictionary definitions, it is useful to distinguish among them in a decision making context.

Criterion. A criterion is a measure of effectiveness of performance. It is the basis for evaluation. Criteria may be further classified as goals or targets and objectives.

Goals. A goal (synonymous with target) is something that is either achieved or not. For example, increasing sales of a product by at least 10% during one year over the previous is a goal. If a goal cannot be or is unlikely to be achieved, it may be converted to an objective.

Objective. An objective is something to be pursued to its fullest. For example, a business may want to maximize its level of profits or maximize the quality of service provided or minimize customer complaints. An objective generally indicates the direction desired.

Attribute. An attribute is a measure that gives a basis for evaluating whether goals have been met or not given a particular decision. It provides a means for evaluating objectives.

Decision Variable. A decision variable is one of the specific decisions made by a decision maker. For example, the planned production of a given product is a decision variable.

1. Complete: They should cover all aspects of a problem.
2. Operational: They can be meaningfully used in the analysis.
3. Decomposable: They can be broken into parts to simplify the process.
4. Nonredundant: They avoid problems of double counting
5. Minimal: The number of attributes should be kept small.

I recommend that at most the magic number of about 7 (see Miller, 1956) objectives be used. Such a limitation tends to keep a problem within the realm of operationality. What happens if there are more than about 7 objectives? First, use constraints to limit outcomes of objectives about which you are sure or about which you feel comfortable about setting such limits. Since constraints must be satisfied at any price, you should not make constraints "too tight." Further, it is useful to check whether feasible alternatives still exist after adding each constraint or after adding a few constraints. An alternative is to treat some of the objectives as goals or targets. We attempt to satisfy the goals. If we can't, we treat them as objectives. We try to get as close to achieving them as possible. We shall go into the idea of doing this mathematically later. Structuring a problem properly is an art, and there is no prescribed way of setting up objectives, goals and constraints.

A Scenario of Management Decision Making

A scenario of management decision making is generally assumed by most academicians:

1. A decision maker (DM) makes a decision.
2. He chooses from a set of possible decisions.
3. The solution he chooses is optimal.

To criticize the scenario, the decision maker, if an individual (as opposed to a group), seldom makes a decision in a vacuum. He is heavily influenced by others. In some instances groups make decisions. Second, the set of possible decisions is not a given. The set of solutions must be generated. The process of determining the set of alternatives may require considerable effort. In many situations, the generation of alternatives was important as the choice mechanism for choosing an alternative.

What is meant by an optimal solution? Since it is impossible to simultaneously maximize all objectives in determining a solution, a more workable definition is needed. A typical definition of optimality is not particularly workable: An optimal decision is one that maximizes a decision maker's utility (or satisfaction). In spite of the limitations of the decision scenario, it is widely used; its limitations are hopefully recognized.

Some Mathematical Considerations of Multiple Criteria Decision Making

The general multiple criteria decision making problem may be formulated as follows:

$$\text{"Maximize"} \quad F(x)$$
$$\text{subject to:} \quad G(x) \leq 0$$

where x is the vector of decision variables, and F(x) is the vector of objectives to be "maximized". In some cases it will be convenient to have an intervening vector y where $F(x) = H(y(x))$. For example, y may be a vector of stochastic objectives which is a function of x. In that case, H would be a vector function of the stochastic objectives.

In some cases F will have some components that are ordinal. Attributes such as quality and convenience of location may only be measurable on an ordinal scale. Further, some objectives may be measured only imperfectly. The word maximize is in quotation marks because maximizing a vector is not a well-defined operation. We shall define it in several ways in what follows.

The constraints $G(x) \leq 0$ are the constraints that define the feasible solution space. They may be stated explicitly and if mathematical be either linear or nonlinear. Alternatively, the alternatives may be stated implicitly by listing them as discrete members of a set. Even though such a set is nonconvex, it is convenient to work with the convex hull of the solutions that generate a convex set.

The formulation of the multiple criteria decision making problem (1) is one that I believe includes virtually all of the approaches developed, as well as the various multiple criteria problems. It is clearly too general, because only very specific forms of problem (1) can be solved optimally in practice using quantitative models. A linear version of problem (1) is as follows:

$$\text{"Maximize"} \quad Cx$$
$$\text{subject to:} \quad Ax \leq b$$

$x_j \geq 0$, if needed, may be included in the constraints $Ax \leq b$.

This particularization of problem (1) is one on which à substantial amount of study has been made. It is referred to as the multiple objective linear programming problem (MOLP) because it is a linear programming problem with multiple objectives. The following theorem is found in several places in the multiple criteria literature.

T h e o r e m: Maximizing a positive weighted ($\lambda > 0$) sum of objectives $\lambda' F (= \sum_i \lambda_i F_i)$ over a set of feasible solutions yields a nondominated solution.

The theorem does not say that for every nondominated solution there exists a set of weights for which the nondominated solution maximizes the weighted sum. As we shall see, that need not be the case.

The second problem that we consider is the discrete alternatives problem. Though it is not a mathematical programming problem, we shall consider it anyway. It is the problem of choosing the best alternative from a discrete set of alternatives. Mathematically, we want to:

$$\text{"Maximize" } F(d)$$
$$\text{subject to: } d \in D$$

where d is a decision and D is the set of possible decisions. Examples of this problem are the purchase of a house or car. For the purchase of a house, D is the set of houses available that satisfy the buyer's constraints. The decision d that is chosen in some sense maximizes the buyer's objectives F(d). With discrete alternatives, even if we are able to conceive of convex combinations of alternatives, we are generally unable to realize them in practice. We cannot blend (or take an average of) a Tudor house at 136 Elmwood Street, and a Ranch house at 3550 Maple Road and come up with a Tuanch house at 1843 (the average) Mapwood Street. Nonetheless, it may still be useful to consider "blends" or convex combinations of discrete alternatives. Further "plotting" discrete alternatives generates a number of discrete points, rather than convex regions of feasible solutions. (The set of convex combinations of a set of discrete solutions is the convex hull of the corresponding solutions points.)

The Objective Functions

Let us now consider the objective functions more carefully. The objective functions may all be assumed to be maximized, without loss of generality, because any objective that is to be minimized can be minimized by maximizing the value of its negative. Accordingly, we shall henceforth refer to objectives to be maximized.

What do we do if we have any goals or targets (as defined earlier)? If they all are simultaneously achievable, we simply add constraints that stipulate the specified value be met and not consider them further. Thus, the achievement of the goals if transformed into an admissible solution satisfying all of the constraints. There is an interesting duality between objectives and constraints, in that the two are closely related.

If the goals are not simultaneously achievable, simply adding constraints as above will lead to no feasible solution to the problem. What must be done in such a situation is to relax some of the goals, or to change goals to objectives as described earlier: to minimize the difference between the goal and the outcome. The idea is to find a solution that is "close" to the goal.

What do we mean by "maximize"? Unlike unidimensional optimization, we want to simultaneously maximize several objectives. Generally that cannot be done. We may define "maximize" in two ways. From a general perspective, one workable definition of "maximize" is to find all nondominated solutions to a problem.

Definition: Dominance

Solution 1 dominates solution 2 if $F(x_1) \geq F(x_2)$ with strict inequality holding for at least one component of F.

A solution is said to be nondominated if no other solution is at least as good as it in every respect and better than it in at least one respect. The concept seems eminently reasonable. By finding all nondominated solutions, one can presumably reduce the number of alternatives. However, for many problems, the number of nondominated alternatives may still be too large to help narrow the choice of alternatives.

It is useful to represent sets of nondominated solutions graphically, particularly when there are two objectives. Consider a set of discrete alternatives for which there are two objectives, both of which are to be maximized. (See Figure 1.) The set of nondominated solutions consist of solutions A, B, D, F, H, J, and K. Solutions C, E, and G are dominated, respectively, by D, F, and H (or J). Were we to consider the set of convex combinations of these solution points, a "blend" of solutions A and D would dominate B, and a "blend" of D and H would dominate F. A nondominated solution, in two dimensions, represented as in Figure 1, has no solutions "northeast" of it.

An example of a a convex set of nondominated solutions may be seen in Figure 4 a few pages ahead.

Figure 1

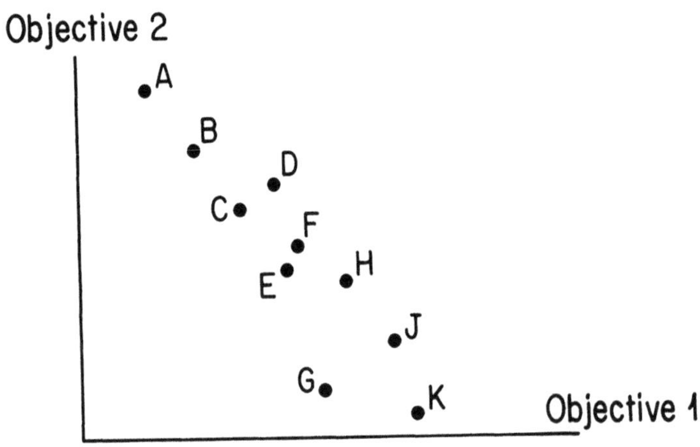

A Discrete Alternatives Example

There may be some instances where we don't want to eliminate dominated solutions. For example, a dominated solution may be sufficiently close to a nondominated solution that we may decide to make a choice based on some secondary criteria not used in the analysis. We may then very well choose the dominated solution based on the secondary criteria. Alternatively, some of the objectives may not be measurable precisely. In such a situation we may not want to exclude dominated alternatives from further analysis. As an example of the first type, suppose a prospective automobile purchaser is choosing among cars on the basis of price, economy, sportiness, and comfort. Suppose further that a foreign-made car appears somehow to be the best choice, but that there is a domestically produced automobile that is its equal in all respects except that the price is slightly higher. The decision maker may nonetheless decide to purchase the domestic automobile because of its better availability of spare parts. In the second instance, suppose that the purchaser is considering two domestically-produced automobiles. We assume as before that the cars are the same for all criteria but one--price. Car A has a lower list price than Car B. However, in the purchase of most automobiles, one can obtain discounts. On haggling with dealers, our purchaser may subsequently find that he can purchase Car B for less than Car A. Hence, if he had excluded Car B because of dominance (on the basis of list price), he would have made a mistake.

The reader may feel that in the first case we should have added spare parts availability to our criteria. Though this could have been done, we may generally use criteria such as this as secondary to resolve close cases. Similarly, it can be argued in the second example that the price variable is transaction price and not list price. Therefore, our selected car is not dominated. Nonetheless, it is difficult to accurately measure transaction price, prior to an automobile purchase transaction.

Another definition of "maximize" is that a decision maker wishes to maximize his utility, as measured by a utility function: a function of the objectives that gives a scalar measure of performance.

Definition: Utility Function

A utility function is a scalar function $u(F(x))$ such that x_1 is preferred to or is indifferent to x_2 if and only if $u(F(x_1)) \geq u(F(x_2))$.

Because of our statement of problem (1), we have at an optimal solution (for any feasible change) the value of the utility function remains the same or decreases. For some multiple criteria methods, we either estimate a utility function u or approximate it locally. In such cases, we use the function or its approximation to identify a most preferred solution.

A Typology of Multiple Criteria Decision Making Models

Quite naturally, different writers have proposed different decision making typologies. My typology consists of two main dimensions:

1. The nature of outcomes--stochastic versus deterministic.
2. The nature of the alternative generating mechanism--whether the constraints limiting the alternatives are explicit or implicit.

These dimensions are indicated in tabular form in Figure 2. The left-hand column includes the implicit constraint models. When the constraints are nonmathematical (implicit or explicit), the alternatives must be explicit. One of a list of alternatives is then selected. The decision analysis problem is included in the implicit constraint category. When the constraints are mathematical and explicit, then the alternative solutions are implicit and may be infinite in number if the solution space is continuous and consists of more than one solution. Problems in the explicit constraint category are generally regarded as mathematical programming involving multiple criteria.

Figure 2

	Implicit Constraints (Explicit Solutions)	Explicit Constraints (Implicit Solutions)
Deterministic Outcomes	Choosing Among Deterministic Discrete Alternatives or Deterministic Decision Analysis	Deterministic Mathematical Programming
Stochastic Outcomes	Stochastic Decision Analysis	Stochastic Mathematical Programming

A Typology of Multiple Criteria Decision Methods

More dimensions may be added to the typology. In addition to implicit constraints versus explicit constraints, and deterministic outcomes versus stochastic outcomes, we can identify other dimensions as well. We may classify the number of decision makers as a dimension: one decision maker versus two decision makers versus three or more decision makers. We may classify the number of objectives, the nature of utility functions considered, as well as the number of solutions found (one solution versus all nondominated solutions). I have chosen only two dimensions because they seem to be the most significant factors for the problems we consider.

In our presentation we consider only deterministic problems: mathematical programming problems and discrete alternative problems. In virtually all of the work on multiple criteria decision making, the spirit of the model employed is not necessarily to determine the best decision (though that is of course desirable!), but to help the decision maker in arriving at his decision. This is what Roy (1977) refers to as "decision aid." It is also what Keeney and Raiffa (1976) refer to as "getting your head straightened out." Before we consider some of the methods in detail, we present two mathematical programming examples. The first is useful in illustrating some concepts; the second will be used in various forms to illustrate the methods. (We earlier gave a graphical representation of a discrete alternative example.)

Two Examples

Consider the following problem, which we shall refer to as Example 1:

$$\begin{array}{lrl}
\text{Maximize} & f_1 = & -x_1 + 2x_2 \\
& f_2 = & 2x_1 - x_2 \\
\text{subject to:} & & x_1 \le 4 \\
& & x_2 \le 4 \\
& & x_1 + x_2 \le 7 \\
& & -x_1 + x_2 \le 3 \\
& & x_1 - x_2 \le 3 \\
& & x_1,\ x_2 \ge 0
\end{array}$$

For this example, x_1 and x_2 are the decision variables, and f_1

and f_2 are the objectives. A plot of the feasible solutions is shown in Figure 3, the maximum solutions indicated (e for f_1, and b for f_2) for each of the objectives. In that figure, we have also identified all of the feasible extreme point solutions as 0 and a through h. In Figure 4 we have plotted the values of the objective functions for this problem. Each of the feasible solutions in Figure 3 has a corresponding point in Figure 4. For example, solution b represented as $x_1 = 4$ and $x_2 = 1$ has objective function values $f_1 = -2$ and $f_2 = 7$ and is so plotted in Figure 3. The nondominated solutions are shown presumably will be found along that line, since any point not on that line is either dominated (below and/or to the left in Figure 4) or infeasible (above and/or to the right in Figure 4).

Figure 3

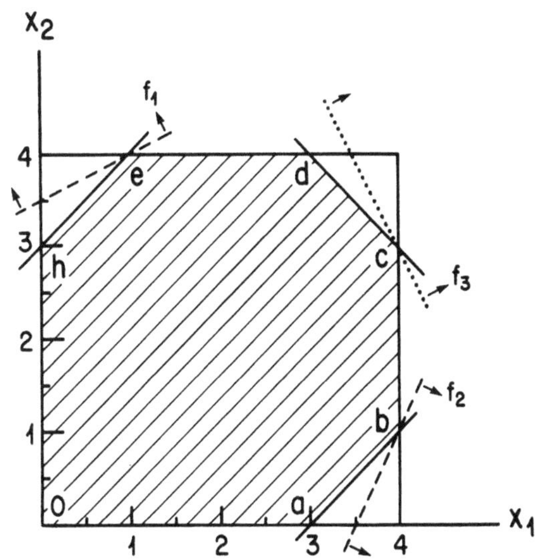

The Feasible Region of Example One and
the Two Objectives f_1 and f_2

Figure 4

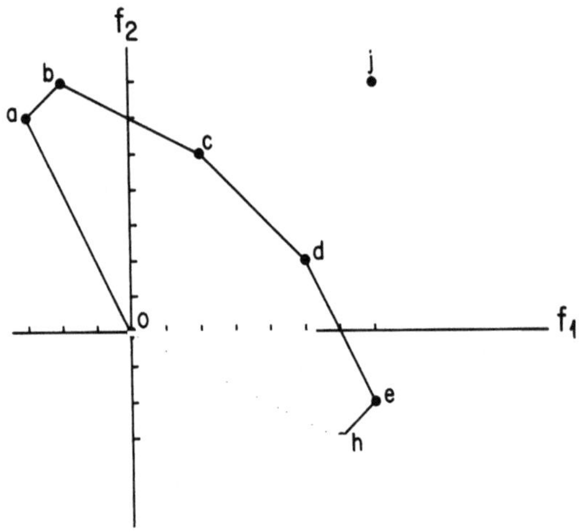

A Plot of the Solutions of the First Example Problem in Objective Function Space: In Terms of the Values of the Objective Function

Since x_1 and x_2 are the decision variables, a graph in terms of x_1 and x_2 (Figure 3) is a graph in decision or activity space. Variables f_1 and f_2 are the objectives; a graph in terms of f_1 and f_2 (Figure 4) is a graph in objective function space. For discrete alternative problems, we cannot normally plot decision variables, because they don't exist. Our example consists of two variables and two objectives. Usually the number of variables is much greater than the number of objectives. We may make the first example more complicated by adding a third objective: $f_3 = 2x_1 + x_2$. See the cross-hatched line in Figure 3. The objective function f_3 is maximized at point c; the plot of the feasible solutions in decision variable space does not change otherwise. To make a plot in objective space with three objectives, we would have to add a third dimension to Figure 4. Rather than do that, we first reconsider Figure 4 with two objectives. Denoting as a

weighted objective function $\lambda_1 f_1 + \lambda_2 f_2$, we can see that (assuming without loss of generality $\lambda_1 + \lambda_2 = 1$ for $\lambda_1 > 2/3$ solution e is optimal. For $\lambda_1 = 2/3$ both solutions d and e (as well as the solutions on the line between them) are optimal. For $1/2 < \lambda_1 < 2/3$, solution d is optimal. Similarly, for $1/3 < \lambda_1 < 1/2$, solution c is optimal, and for $0 < \lambda_1 < 1/3$, solution b is optimal. Because $\lambda_2 = 1 - \lambda_1$, we could plot the regions along a straight line. Adding a third objective gives us a weighted objective function $\lambda_1 f_1 + \lambda_2 f_2 + \lambda_3 f_3$. Now using the restriction $\lambda_1 + \lambda_2 + \lambda_3 = 1$ or $\lambda_3 = 1 - \lambda_1 - \lambda_2$ and eliminating λ_3 we may draw the regions in which each solution is optimal. We call the corresponding space the weight space.
See Figure 5 where we read off the values of λ_1 and λ_2 and compute $\lambda_3 = 1 - \lambda_1 - \lambda_2$. The solutions with $\lambda_3 = 0$ (i.e., $\lambda_1 + \lambda_2 = 1$) are still valid; they appear along the line $\lambda_1 + \lambda_2 = 1$. Other solutions are indicated accordingly.

Figure 5

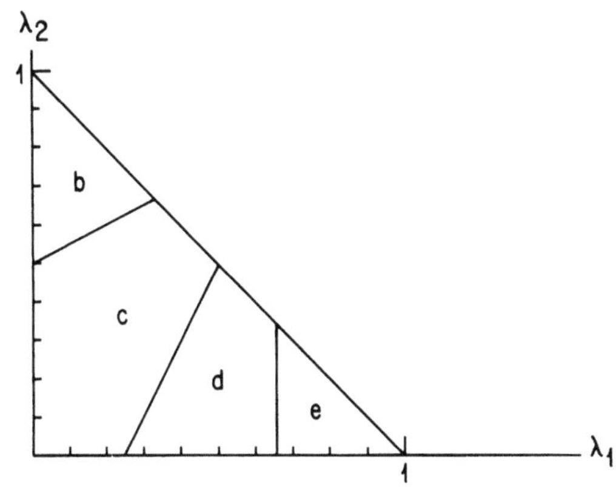

A Plot Indicating the Values of λ_1 and λ_2 ($\lambda_3 = 1 - \lambda_1 - \lambda_2$)

We now consider a more complicated example, Example 2.

$$2x_1 + x_2 + 4x_3 + 3x_4 \leq 60 \qquad \text{(slack } x_5\text{)}$$
$$3x_1 + 4x_2 + x_3 + 2x_4 \leq 60 \qquad \text{(slack } x_6\text{)}$$
$$x_1, x_2, x_3, x_4 \geq 0$$

Three objectives are to be maximized:

$$u_1 = 3x_1 + x_2 + 2x_3 + x_4$$
$$u_2 = x_1 - x_2 + 2x_3 + 4x_4$$
$$u_3 = -x_1 + 5x_2 + x_3 + 2x_4$$

The problem has nine basic feasible solutions. The values of the decision variables and the objective functions are given below (all omitted variables are zero):

1.	$x_1 = 18$,	$x_3 = 6$,	$u_1 = 66$,	$u_2 = 30$,	$u_3 = -12$
2.	$x_4 = 20$,	$x_6 = 20$,	$u_1 = 20$,	$u_2 = 80$,	$u_3 = 40$
3.	$x_2 = 15$,	$x_5 = 45$,	$u_1 = 15$,	$u_2 = -15$,	$u_3 = 75$
4.	$x_2 = 6$,	$x_4 = 18$,	$u_1 = 24$,	$u_2 = 66$,	$u_3 = 66$
5.	$x_1 = 12$,	$x_4 = 12$,	$u_1 = 48$,	$u_2 = 60$,	$u_3 = 12$
6.	$x_2 = 12$,	$x_3 = 12$,	$u_1 = 36$,	$u_2 = 12$,	$u_3 = 72$
7.	$x_3 = 15$,	$x_6 = 45$,	$u_1 = 30$,	$u_2 = 30$,	$u_3 = 15$
8.	$x_1 = 20$,	$x_5 = 20$,	$u_1 = 60$,	$u_2 = 20$,	$u_3 = -20$
9.	$x_5 = 60$,	$x_6 = 60$,	$u_1 = 0$,	$u_2 = 0$,	$u_3 = 0$

The first six solutions are nondominated, the last three are dominated. Figure 6 indicates which solutions are adjacent extreme point solutions of which other solutions (i.e., they differ by precisely one basic variable).

In order to plot the problem solutions in activity space we need to plot a 4-dimensional graph! More reasonable is plotting the objectives in three dimensions. However, instead we present the plot for the weights λ_1 and λ_2 (and λ_3) as we did for Example 1. See Figure 7.

Any solutions which have a common edge in Figure 7 are adjacent. (See Figure 6.) However, some solutions are adjacent (e.g., 3 and 4), yet do not have a common edge.

Figure 6

Solution	is adjacent to Solutions
1	5, 6, 7, 8
2	4, 5, 7, 9
3	4, 6, 8, 9
4	2, 3, 5, 6
5	1, 2, 4, 8
6	1, 3, 4, 7
7	1, 2, 6, 9
8	1, 3, 5, 9
9	2, 3, 7, 8

Adjacency of Basic Feasible Solutions of Example 2

Figure 7

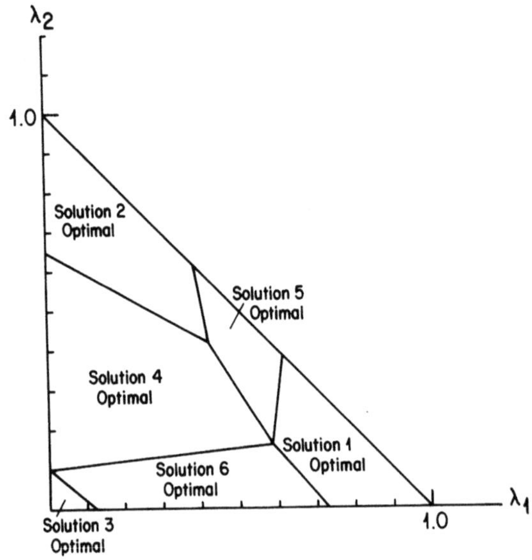

A Plot of λ Values and the Corresponding Optimal Solutions

For discrete alternative problems, we may plot the points, as we did in Figure 1, in the objective function space. If we consider the convex hull (set of convex combinations) of all such points, we may also construct a weight space for the problem. We cannot, as we indicated earlier, construct a decision space for such problems.

Some Naive Methods of Solving Multiple Criteria Mathematical Programming Problems

There are several naive methods for solving multiple criteria mathematical programming problems. They are simple in concept, though generally not very good. They were early approaches to the problems, and have evolved into current approaches. We shall first consider multiple criteria linear programming problems.

1. Setting Levels of All Objectives

The first of the naive methods to be considered is that of specifying or setting levels of all objectives, and then solving for a feasible solution. The approach is to specify a vector of objectives d such that $Cx = d$. The object then is to find a feasible solution to the set of constraints:

$$Cx = d$$

$$Ax \leq b,\ x \geq 0$$

The problem can be solved as a linear programming problem, and there are three possible outcomes as illustrated in Figure 8 for a two-objective problem. The feasible region is indicated. The three possible outcomes are as follows:

a. No feasible solution

b. A dominated solution

c. A nondominated solution.

These are illustrated in Figure 8. If the objectives are set too high, there is no feasible solution (e.g., point a). If the objectives are not set high enough, a feasible solution that is dominated (e.g., solution b) will be found. Almost certainly one of these two outcomes will occur. Only in rare circumstances would simply selecting a vector d yield an efficient (or nondominated) solution. Given two points such as a and b, we can sometimes use a line search for a nondominated solution on the line segment connecting them (e.g., line segment ab; the nondominated solution on the line segment between them is point e). That this does not necessarily happen is illustrated by feasible point k and infeasible point h; there is no efficient point on the line segment joining them.

Even if we had a method for finding all efficient solutions, we would still not necessarily know which solution is best. Methods that set levels of all objectives but overcome some of the limitations include goal programming and a method that has been developed by Wierzbicki (1980). These are discussed in later sections. See also the step method (Benayoun et al. (1971).

Figure 8

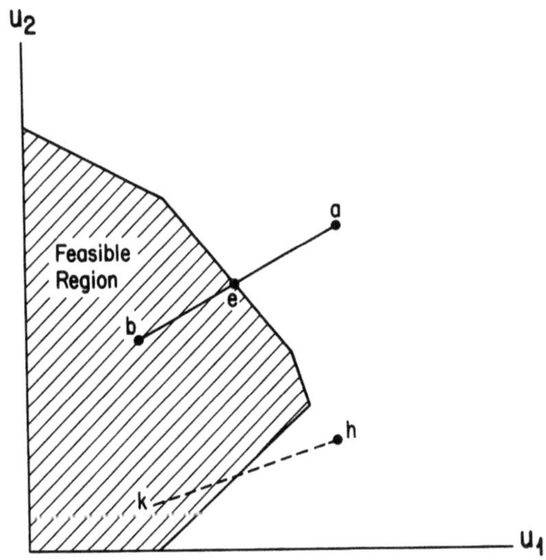

A Graph of a Simple Two Dimensional Example

2. Setting Minimal Levels of All But One Objective

A second naive approach is to set minimum levels for all but one objective and to maximize the remaining objective.

Mathematically this amounts to solving a linear programming problem of the following form:

$$\begin{aligned} \text{Maximize} \quad & c_1x \\ \text{subject to} \quad & c_2x \geq d_2 \\ & c_3x \geq d_3 \\ & c_px \geq d_p \\ & Ax \leq b, \quad x \geq 0 \end{aligned}$$

where $d_2,\ldots.,d_p$ are the minimum levels of objectives $2,\ldots,p$ and $c_1,c_2,\ldots.,c_p$ are the p objective function vectors. We have chosen to maximize the first objective without loss of generality. The result will certainly be a nondominated solution*. For our example problem of Figure 3 there are infinitely many solutions along the line segments bc, cd, and de. Presumably, one (or more) of these solutions is preferred to the others. Which of these solutions is most preferred by the decision maker? That is not clear. A method that employs this approach has been developed by Haimes and Hall (1974.)

3. Finding All Efficient Extreme Point Solutions

Multiple Objective Linear Programming (MOLP) to find all nondominated or efficient solutions has been widely proposed as another approach. The concept of vector maximum and its early consideration by researchers (see for example, Charnes and Cooper, 1961) has been around for a long time. Only in the early 70s was it considered seriously as a computational procedure.

---

*Ralph Steuer has pointed out that solutions to such problems may be in some cases weakly dominated, though such solutions may be avoided.

Evans and Steuer (1973) and Yu and Zeleny (1975) generated and solved problems of several sizes to obtain all nondominated extreme point solutions. The results were not good, except for two-objective problems for which parametric programming may be used. Basically, the methods consider the linear programming problem:

$$\text{Maximize } \lambda'Cx$$

$$\text{subject to: } Ax = b$$

$$x \geq 0$$

where the vector of weights $\lambda > 0$. For every nondominated extreme point solution, there exists a convex cone in space, that is a cone for which $\lambda'(C_N - C_B^{-1}N) \leq 0$, using the usual linear programming notation (C_N and C_B are the nonbasic and basic partitions, respectively, of C. N is the complement of B with respect to A.) The λ-space shown in the various figures of the text is the intersection of all cones with the constraint $\sum\lambda_j = 1$. The methods for finding all nondominated extreme point solutions essentially enumerate the convex cones. The idea was that all efficient solutions could be computed, and the decision maker could choose from them. Since there are in general far too many, the approach is not workable in practice. Steuer's contracting cone method, described in a later section, partially overcomes the problems.

4. Using Weights to Combine Objective Functions

The idea of using weights seems to be an attractive one. It involves averaging or blending the objectives into a composite objective and then maximizing the result. The difficulty is in specifying weights. It is incorrect to say that if the weight for one objective is larger than that of another, that the first objective is more important than the second and vice versa. The weights depend upon the units in which the objectives are measured. For example, equal weights have a rather unequal effect if objectives are to maximize GNP measured in billions of dollars and to maximize the fraction of the population who are above the poverty level as measured by a number between zero and one. The second objective will in that case have virtually no effect. The Zionts-Wallenius method (considered below) extends and uses this approach.

Overcoming the Problems of the Naive Approaches

Several of the naive approaches have appealing characteristics, which no doubt led to their development. To overcome some of the problems with the methods, further development was done on these methods. We now describe the results.

1. Goal Programming

The concept of goal programming, effectively a method for setting all objectives, was introduced by Charnes and Cooper (1961) and extended by Ijiri (1965) and Lee (1972), among others. Goal programming involves the solution of linear programming problems (although other mathematical programming forms such as integer programming have also been formulated in a goal programming context) with several goals or targets. Generally, goal programming assumes a linear constraint set of the (matrix) form $Ax = b$, $x \geq 0$ where x is the vector of decision variables. Denoting an objective as $c_i x$, there are several possible forms, all of which can be rewritten as $h_i \leq c_i' \leq u_i$ where h_i is the desired lower bound on objective i, and u_i is the desired upper bound. The bound constraints are not "hard" in that they may be violated. First add variables s_i and t_i and rewrite the bound constraints as

$$c_i'x - s_i \leq u_i \,, \qquad i = 1,\ldots,p$$
$$c_i'x + t_i \geq h_i \,, \qquad i = 1,\ldots,p$$

where p is the number of objectives or goals. Now using matrix notation with $c_i = (c_{i1}mc_{i2},\ldots,c_{ip})'$, $s = (s_1,s_2,\ldots,s_p)'$, $t = (t_1,t_2,\ldots,t_p)'$, $k = (k_1,k_2,\ldots,k_p)'$, $q = (q_1,q_2,\ldots,q_p)'$, $u = (u_1,u_2,\ldots,u_p)'$ and $h = (h_1,h_2,\ldots,h_p)'$ we wish to

$$\begin{array}{ll} \text{Minimize} & k's + q't \\ \text{subject to:} & Cx - s \leq u \\ & Cx + t \leq h \\ & Ax = b \\ & x, s, t \geq 0 \end{array}$$

where k and q are vectors of weights to measure the violations of the bound constraints. If desired, several different s and t variables may be used for each goal with different values of k and q as well as upper bounds on the s and t variables. The effect of this is to allow for piece-wise linear, nonlinear penalties in the failure to achieve goals. As outlined, the relationships yield convex sets. For more information on these nonlinearities and others as well as nonconvex nonlinearities, see Charnes and Cooper (1977).

Instead of minimizing a weighted sum of deviations from goals, goal programming may be used to minimize the maximum deviation from a set of goals. This is done by changing the formulation by adding the constraints

$$q_i s_i \leq z$$
$$k_i t_i \leq z$$

and changing the objective to minimize z. The effective objective then is to minimize $(\max\{q_i s_i, k_i t_i\})$, the maximum weighted deviation from a goal.

Another variation of goal programming employs preemptive priorities instead of numerical weights. Let some subset of weights have much greater values than another subset of weights so that any finite multiple of the weights of the latter set is always less than any of the weights of the former set. The effect is to first minimize the weighted sums for the highest preemptive priority group. Then constraining that weighted sum to be equal to its minimum value, the next highest preemptive priority group sum is minimized, and so on, for as many preemptive priority groups as there may be.

Where goal programming falls flat is in the selection of the goals as well as the specification of the weights, that is the vectors k and q. The selection of goals should not be a difficult problem, although it is important for the decision maker to be aware of tradeoffs which face him. The weights must be selected by the user, and goal programming does not have much to say about the choice of weights. About the only device that is offered in terms of weights is preemptive priorities, which we have already considered. Nonetheless, goal programming has been fairly widely used in practice because of the ease of specifying a goal vector, and the ease of understanding what is going on. We consider Example 2 as a goal programming problem, where our objective is to minimize the absolute sum of deviations below the goal values (66, 80, 75). For example, the solution (60, 75, 75) would have a value of $|66-60| + |80-75| + |75-75|$ or 11.

The formulation is as follows:

$$\begin{array}{llr}
\text{Minimize} & t_1 + t_2 + t_3 & \\
\text{subject to:} & 3x_1 + x_2 + 2x_3 + x_4 + t_1 & \geq 66 \\
& x_1 - x_2 + 2x_3 + 4x_4 + t_2 & \geq 80 \\
& -x_1 + 5x_2 + x_3 + 2x_4 + t_3 & \geq 75 \\
& 2x_1 + x_2 + 4x_3 + 3x_4 & \leq 60 \\
& 3x_1 + 4x_2 + x_3 + 2x_4 & \leq 60 \\
& x_1, x_2, x_3, x_4, t_1, t_2, t_3 & \geq 0
\end{array}$$

The optimal solution to the above problem is $x_2 = 6$, $x_4 = 18$, $t_1 = 42$, $t_2 = 14$, $t_3 = 9$ (all other variables are zero), or $u_1 = 24$, $u_2 = 66$, $u_3 = 66$. Changing the objective function to minimize $3t_1 + t_2 + t_3$ changes the solution to $u_1 = 66$, $u_2 = 30$, $u_3 = -12$, to illustrate another set of weights.

If we now add to the formulation $t_1 \leq z$, $t_2 \leq z$, and $t_3 \leq z$ and change the objective to minimize z we have an example of minimizing the maximum deviation from each goal. We obtain the solution $x_1 = 4.86$, $x_2 = 5.01$, $x_3 = 2.88$, $x_4 = 11.24$, $t_1 = t_2 = t_3 = t_4 = z = 29.40$, or $u_1 = 36.6$, $u_2 = 50.6$, and $u_3 = 45.6$.

We now illustrate the use of preemptive priorities. Let us assume that our first priority is to get u_1 to 50, our second priority is to get u_2 to 50, and our third priority is to get u_3 to 50.

We formulate the problem as follows:

$$
\begin{array}{lr}
\text{Minimize} \quad P_1t_1 + P_2t_2 + P_3t_3 & (P_1 >> P_2 >> P_3) \\
3x_1 + x_2 + 2x_3 + x_4 + t_1 \geq 50 & (u_1 \geq 50) \\
x_1 - x_2 + 2x_3 + 4x_4 + t_2 \geq 50 & (u_1 \geq 50) \\
-x_1 + 5x_2 + x_3 + 2x_4 + t_3 \geq 50 & (u_2 \geq 50) \\
2x_1 + x_2 + 4x_3 + 3x_4 \leq 60 & \\
3x_1 + 4x_2 + x_3 + 2x_4 \leq 60 & \\
x_1, \ldots, x_4, t_1, t_2, t_3 \geq 0 &
\end{array}
$$

(>> means much greater than)

The optimal solution to the problem is $u_1 = 50$, $u_2 = 50$, $u_3 = 13.65$ ($x_1 = 11.88$, $x_2 = 1.18$, $x_3 - 2.24$, and $x_4 = 8.71$). The procedure is that we first minimize P_1t_1. When we achieve its minimum (here zero), we fix P_1t_1 at that value and minimize P_2t_2. Once we have found that minimum, we fix both P_1t_1 and P_2t_2 at their minima, and minimize P_3t_3, and so on.

2. Scalarizing Functions and the Method of Wierzbicki

Wierzbicki (1980) has developed a method which may be thought of as a method for setting levels of all objectives. It assumes that all objectives are to be maximized, and employs a scalarizing function to find an efficient solution. Referring to our naive version, the chosen levels of objectives are almost certainly infeasible or dominated. The scalarizing method or reference point approach, as it is also called (see Kallio,

Lewandowski, and Orchard-Hays, 1980) finds the closest efficient solution to the chosen point. It is intended to be used in a simulation-type mode by the decision maker.

Although there are a wide variety of scalarization functions that could be used, one that seems quite effective is one which can be represented in a linear programming context. Let u_i be the target level for objective i. The objective is to maximize

$$\left\{ \min \left\{ \rho \min \{C_i x - u_i\} , \sum_i (C_i x - u_i) \right\} + \epsilon(C_i x - u_i) \right\}$$

where the parameter $\rho \geq p$ the number of objectives and ϵ is a non-negative vector of small positive numbers. This objective function is achieved by a similar representation to that in goal programming for the minimization of the maximum deviation from a set of goals. Here, however, we maximize the minimum of (1) a constant times the minimum overachievement of a goal, and (2) the sum of overachievements of goals, averaged together with a weighted overachievement of goals. As in the case of goal programming, the function and parameters are somewhat arbitrary. However, the purpose of this method is to be used as an efficient solution generator, one that can be used to generate a sequence of efficient solution points. It is rather similar to goal programming and has been programmed to solve problems having as many as 99 objectives with as many as 1000 constraints.

3. Steuer's Contracting Cone Method

Steuer's Contracting Cone Method (Steuer and Schuler (1976) and Steuer and Wallace (1978), see also Steuer (1977 and 1986)) is a refinement to the generation of all nondominated solutions that generates only a relatively small number of nondominated extreme point solutions. It does this by selecting a convex cone in λ space that is large initially and includes sets of weights corresponding to many nondominated extreme point solutions. Rather than generating all of them, however, he generates only a very small number of extreme point solution,and questions the decision maker regarding their relative attractiveness. He then uses the responses to contract the cone. When the cone becomes sufficiently small, the method generates all of the nondominated extreme point solutions in the cone for final consideration by the decision maker.

Assuming that there are p objectives, Steuer's method generates 2p + 1 trial solutions each time. The vectors generated* are:

	Values in General	Initial Values
λ_1	= the first extreme vector	(1,0,0,...,0)
λ_2	= the second extreme vector	(0,1,0,...,0)
.		
.		
λ_p	= the pth extreme vector	(0,0,0,...,1)

$\lambda_{p+1} = 1/p\ (\lambda_1 + \lambda_2 + \ldots + \lambda_p) = (1/p, 1/p, 1/p, \ldots, 1/p$

$\lambda_{p+2} = (\lambda_2 + \lambda_3 + \ldots + \lambda_{p+1})/p = (1/p^2, r, r, r, \ldots, r)$

$\lambda_{p+3} = (\lambda_1 + \lambda_3 + \lambda_4 + .. + \lambda_p + \lambda_{p+1})/p = (r, 1/p^2, r, r, \ldots, r)$

$\lambda_{p+4} = (\lambda_1 + \lambda_2 + \lambda_4 + .. + \lambda_p + \lambda_{p+1})/p = (r, r, 1/p^2, r, \ldots, r)$

.

.

$\lambda_{2p+1} = (\lambda_1 + \lambda_2 + \ldots + \lambda_{p-1} + \lambda_{p+1})/p = (r, r, r, \ldots, r, 1/p^2)$

where $r = (p + 1)/p^2$

---

*Instead of using zeros in the vector, we use some sufficiently small positive numbers. However, for simplicity of presentation we use zeros here.

The first p vectors are the extreme vectors of the cone, the p + 1st is the mean or center of gravity of the first p vectors, and each of the others is the mean or center of gravity of p-1 extreme vectors and the p + 1st vector. For each of the weights λ, a linear programming problem is solved maximizing $\lambda' Cx$, and the decision maker is presented with the 2p + 1 solutions. (All these solutions may not be different.) He is asked to choose which of the solutions he likes most, or if he is ready to look at all of the extreme point solutions in the cone. In the latter case, all extreme point solutions in the cone are found and presented to the decision maker for a final choice. Otherwise, the cone is contracted about the selected extreme point solution. The first p vectors (the extreme vectors) for the next iteration are the vectors corresponding to the chosen solution, say λ_q, and the average of that vector with each of the (first) p extreme vectors from the previous iteration.

$$\lambda'_i(i=1,\ldots,p) = .5\lambda_i + .5\lambda_q$$

The prime indicates the new trial weights. The remaining p + 1 vectors are found from $\lambda'_1,\ldots,\lambda'_p$ as λ_{p+i}, i =

Figure 9

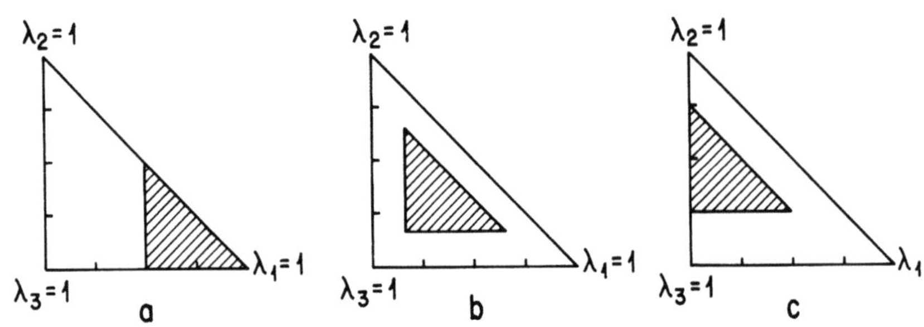

An Illustration of the Three Cases of Contracting Cones in Weight Space for the Case of Three Objectives

$1,\ldots,p+1$ are found from $\lambda_1,\ldots,\lambda_p$.

(The larger cone is the cone generated by the previous set of weights, and the smaller or shaded cone is the cone generated by the new set of weights)

Case a The solution corresponding to $\lambda_1 = 1$, $\lambda_j = 0$ $(j \neq 1)$, is preferred.

Case b The solution corresponding to $\lambda_j = 1/p$, $j = 1,\ldots,p$, is preferred.

Case c The solution corresponding to $\lambda_1 = 1/p^2$, $\lambda_j = (p+1)/p^2$, $j \neq 1$ is preferred.

The process is repeated until the decision maker asks for all efficient solutions defined in a cone to make a final decision. The effect of contracting the cone is to reduce the volume of the cone to $(1/2)^p$ of what it was prior to the contraction. This fraction could be adjusted to a larger or smaller fraction, if desired. To illustrate how the cone contracts as we've described, consider a three objective problem. Figure 9 illustrates such a cone section with $\lambda_1 + \lambda_2 + \lambda_3 = 1$. If we contract the cone about one of the original p extreme vectors (e.g., λ_1), we have the diagram shown in Figure 9a. If we contract the cone about the center of gravity (the mean -- λ_4), we have the diagram shown in Figure 9b. Finally, if we contract the cone about one of the off-center solutions (e.g., λ_5), we have the diagram shown in Figure 9c.

The procedure is appealing, but heuristic in nature. It does not necessarily always find the optimal solution. However, regarding performance, Steuer and Schuler (1976) report favorable experience in applications to a forestry management problem.

4. The Zionts-Wallenius Method

The Zionts-Wallenius (1976, 1983) method for multiple objective linear programming uses weights. In that framework a numerical weight (arbitrary initially) is chosen for each objective. Then each objective is multiplied by its weight, and all of the weighted objectives are then summed. The resulting composite objective is a proxy for a utility function. (The manager need not be aware of the combination process.) Using the composite objective, solve the corresponding linear programming problem. The solution to that problem, an efficient solution, is presented to the decision maker in terms of the levels of each objective achieved. Then the decision maker is offered some trades from that solution, again only in terms of the marginal changes to the objectives. The trades take the form, "Are you willing to reduce objective 1 by so much in return for an increase in objective 2 by a certain amount, an increase in objective 3 by a certain amount, and so on?" The decision maker is asked to respond either yes, no, or I don't know to the proposed trade. The method then develops a new set of weights consistent with the responses obtained, and a corresponding new solution. The process is then repeated, until a best solution is found.

The above version of the method is valid for linear utility functions. However, the method is extended to allow for the maximization of a general but unspecified concave function of objectives. The changes to the method from that described above are modest. First, where possible the trades are presented in terms of scenarios, e.g., "Which do you prefer, alternative A or alternative B?" Second, each new nondominated extreme point solution to the problem is compared with the old, and either the new solution, or one preferred to the old one is used for the next iteration. Finally, the procedure terminates with a neighborhood that contains the optimal solution. Experience with the method has been good. With as many as seven objectives on moderate-sized linear programming problems (about 300 constraints) the maximum number of solutions is about ten, and the maximum number of questions is under 100.

We describe the general concave (GC) version in more detail. The linear problem form may, of course, be solved as a special case, though the GC method does not reduce to the linear method in that case.

We repeat the formulation of the problem for convenience.

$$\text{Maximize} \quad g(Cx)$$
$$\text{subject to:} \quad Ax \leq b, \quad x \geq 0$$

The underlying concave utility function g is assumed to have continuous first derivatives. We present the algorithm as a sequence of steps.

1. Choose an arbitrary vector of weights, $\lambda > 0$.
2. Solve the linear programming problem

$$\text{Maximize} \quad \lambda' Cx$$

$$\text{subject to:} \quad Ax \leq b, \quad x \geq 0$$

The result is a nondominated extreme point solution x^*. If this is the first time through this step, go to step 3. Otherwise, ask whether solution x^* is preferred to the old x^* solution. If yes, discard the old solution and go to step 3. If no, replace x^* by x^o and go to step 3.

3. Find all adjacent efficient extreme point solutions to x^* consistent with prior responses. If there are none, drop the oldest set of responses and repeat step 3. Otherwise go to step 4.

4. (This step is simplified over what is used. See Zionts and Wallenius (1983).) Ask the decision maker (DM) to choose between x^* and an adjacent efficient extreme point solution. Do not repeat any questions previously asked. If the objective function values of the solutions are too close, or if x^* was preferred to an adjacent solution, ask the decision maker about the tradeoffs between the two solutions. The DM may indicate which solution he prefers, or indicate that he cannot choose between the two. If he prefers no alternatives or tradeoffs go to step 5. Otherwise mark a solution preferred to x^* as x^o and go to step 6.

5. If all previous responses have been deleted, stop; if the decision maker does not like any tradeoffs from x^*, the optimal solution is x^*. Otherwise, to find the optimal solution, the method terminates and a search method (not part of this method) must be used to search the facets. If the procedure does not stop in which case previous responses have not been deleted, delete the oldest set of responses and go to step 3.

6. Find a set of weights $\lambda > 0$ consistent with all previous responses. If there is no feasible set, delete the oldest response and repeat step 6. When a feasible set of weights is found, go to step 2.

To find the adjacent efficient extreme points in step 3, consider the tradeoffs offered $(w_{1j},\ldots,w_{pj})$ by moving to the adjacent extreme point solution j. Then, consider the following linear programming problem:

$$\text{Maximize} \quad \sum_{i=1}^{p} w_{ik}\lambda_i$$

$$\text{subject to:} \quad \sum_{i=1}^{p} w_{ij}\lambda_i \le 0 \quad j \in N,\ j \ne k \qquad \text{(A)}$$

$$\lambda_i \ge 0 \qquad i = 1,\ldots,p$$

where N is the set of nonbasic variables corresponding to solution on x^*. No convex combination of tradeoffs dominates the null vector, for otherwise solutions x^* would not be efficient.

<u>Definition</u>: Given two efficient extreme point solutions x_a and x^*, solution x_a is an adjacent efficient extreme point solution of x^* if and only if all convex combinations of x^* and x_a are efficient solutions.

<u>Theorem</u>: The optimal solution to problem (A) is zero if and only if solution k offering the tradeoff vector $w_{1k},\ldots,w_{pk}$ is not an efficient vector of the set of vectors w_j, $j \in N$. (For a proof see Zionts and Wallenius (1976).)

<u>Corollary</u>: If problem (A) has a positive infinite solution, then solution k offering the tradeoff vector $w_{1k},\ldots,w_{pk}$ is an efficient vector of the set of vectors w_j, $j \in N$.

The method does not explicitly solve problem (A) for every value of k. What it does is to choose one value of k at a time and to solve (A) for that value of k. At each iteration, a sequence of tests for other values of k are made which in general eliminate solving problems for other values of k.

As an example of the method, we consider Example 2. As a, "true" set of weights we use $\lambda_1 = .58$, $\lambda_2 = .21$, and $\lambda_3 = .21$. Our solution procedure begins with $\lambda_1 = \lambda_2 = \lambda_3 = 1/3$. Refer to Figure 7 for further insight. The initial solution is solution 4. First the decision maker is asked to compare solutions 4 and 2, he should prefer 4. Considering 4 versus 5, he should prefer 5. Considering 4 versus 6, he should prefer 4. A consistent set of weights is $\lambda_1 = .818$, $\lambda_2 = .182$, $\lambda_3 = 0$, and the new solution* is solution 1. The decision maker is asked to choose between 1 and 5; he should prefer 5. A set of consistent weights is .594, .160, .246. They yield solution 5. A final question is asked: between 5 and 2. Since he should prefer 5, there are no further questions to ask; solution 5 is optimal.

---

* We don't use zero weights: λ_3 would be equal to some sufficiently small positive number.

The Zionts-Wallenius method is extended to integer programming in Zionts (1977), which is extended, implemented, and tested in Villareal (1979) and Ramesh (1985). See also Karwan, Zionts, Villareal, and Ramesh (1985).

The Zionts-Wallenius method has been used by several organizations and has met with success. For example, Wallenius, Wallenius, and Vartia (1978) describe an application to macroeconomic planning for the Government of Finland. They used an input-output model of the Finnish economy with four objectives chosen by the Finnish Economic Council chaired by the Prime Minister. The objectives were:

1. the percentage change in gross domestic produce
2. unemployment
3. the rate of inflation as measured by consumer prices
4. the balance of trade.

They first tried using the Geoffrion, Dyer, and Feinberg (1972) approach using an improvement prescribed by Dyer (1973). Although the method worked, the users found the estimation of the marginal rates of substitution difficult. Then the Zionts-Wallenius method was used. Results were obtained that were quite satisfactory.

One criticism of the Zionts-Wallenius approach is that if the underlying utility function is nonlinear, at termination we may not always have an optimal solution. (For an underlying linear utility function, the procedure gives an optimal solution.) However, the termination of the procedure indicates when this does occur. In such instances, we will have an extreme point solution that is preferred to all adjacent efficient extreme point solutions. A search procedure will then have to be used to find the optimal. See, for example, Deshpande (1980).

5. The Geoffrion, Dyer, and Feinberg Method

The next mathematical programming method to be discussed, that of Geoffrion, Dyer, and Feinberg (1972), is in the spirit of a weighting method. However, it is a gradient type of method that allows for a nonlinear problem. The method begins with a decision that satisfies all of the constraints. Then information is elicited from the decision maker indicating how he would like to alter the initial levels of the various objectives. More specifically, he is asked to indicate how much of a reference criterion he is willing to give up in order to gain a fixed amount on one of the other criteria. The responses are elicited for every criterion except the reference criterion.

To illustrate, suppose that one has three objectives:

1. to maximize return on investment;
2. to maximize growth in sales;
3. to minimize borrowing.

Given a starting feasible solution and taking return on investment as our reference criterion, the decision maker would be asked two questions to consider from that solution:

1. What percentage growth in sales must you gain in order to give up a 1% return on investment?
2. What decrease in borrowing must you achieve in order to give up a 1% return on investment?

His responses can be used to determine the direction of change in objectives most desired.

That direction is then used as an objective function to be maximized, and the solution (the new solution) maximizing the objective is found. Then a one-dimensional search is conducted with the decision maker from the previous solution to the new solution. The decision maker is asked in a systematic manner to choose the best decision along that direction. Using the best decision as a new starting point, a new direction is elicited from the decision maker as above and the process is repeated until the decision maker is satisfied with the solution.

We now give an example of the Geoffrion, Dyer, and Feinberg method. If we were to assume a linear utility function, and provide correct tradeoff information, the method requires only one iteration. If we assume a nonlinear utility function or consider a linear utility function and do not provide correct tradeoff information, more iterations are required. We choose to assume a nonlinear utility function. We use example two; our utility function is

$$\text{Maximize} \quad U = -(u_1 - 66)^2 - (u_2 - 80)^2 - (u_3 - 75)^2$$

The constraints are as before.

We start with the solution $u_1 = u_2 = u_3 = x_1 = x_2 = x_3 = x_4 = 0$ ("true" objective function value -16,381). The partial derivatives are

$$\frac{\partial U}{\partial u_1} = -2(u_1 - 66), \quad \frac{\partial U}{\partial u_2} = -2(u_2 - 80), \text{ and } \frac{\partial U}{\partial u_3} = -2(u_3 - 75).$$

For the initial solution the vector of partial derivatives is 132, 160, 150, normalized as .299, .362, .339. We solve the linear programming problem using this set of weights to combine objectives. From Figure 7 (p. 17) we see the solution with that set of weights is solution 4 (24 66 66). We then choose the best solution on the line segment between (0 0 0) and (24 66 66) which is (24 66 66) (with true objective function value -2041). We find this by searching along the line segment between the two solutions. The new normalized objective function vector at (24 66 66) is .646 .215 .138, and the solution for that set of weights is solution 5 (48 60 12). The maximum solution is (26.6 63.8 61.2) on the line segment between the two solutions (See Table 1). The new normalized objective function vector at that point is .646 .215 .138, and the maximizing solution for that set of weights is solution 5 (48 60 12). The maximum solution on the line segment between (26.6 63.8 61.2) and (48 60 12) is (27.4 63.7 59.5) (with true objective function value -1999). At this point we begin to alternate between maximizing solutions four and five until the solution converges. The first few solutions and the optimum are summarized in Table 1. The optimal solution is approximately $x_1 = 1.5$, $x_2 = 5.25$, $x_4 = 17.25$ with objective function values (27.0 65.25 59.25).

An application of the method to the operation of an academic department on a university campus is described based on data from the 1970-1971 operations of the Graduate School of Management, University of California, Los Angeles. A linear programming model of the problem was developed and used to formulate annual departmental operating plans. Six criteria for evaluation were stated, including number of course sections offered at various levels, the level of teaching assistance used, and faculty involvement in various nonteaching activities. The decision variables under the control of the department were the number of course sections offered at different levels, the number of regular and temporary faculty hired, and the number of faculty released from teaching. The starting point for the analyst was the previous year's operating position, and the resulting solution suggested an important reallocation of faculty effort from teaching to other activities. The method was used without significant difficulty, and the results were adopted by the department, according to the article.

TABLE 1

	Solution			Objective Function Value	Maximizing Solution		
1	0	0	0	-16,381	24	66	66
2	24	66	66	- 2,041	66	30	-12
3	26.6	63.8	61.2	- 2,005	48	60	66
4	27.4	63.7	59.5	- 1,999	24	66	66
5	27.0	64.0	60.3	- 1,993	48	60	12
6	27.34	63.94	59.52	- 2,992.15	24	66	66
7	27.092	64.093	60.002	- 1,986.6	48	60	12
.							
.							
.							
.							
∞	27.0	65.25	59.25	- 1,986.6	24	66	66
					48	60	12

The problems with the method are asking the user to evaluate the gradient and asking the user to choose a solution along a line segment in the search procedure.

Some Hybrid Approaches

We next consider two hybrid methods, methods that include one or more of the ideas of the naive approaches in their implementation.

6. The Visual Interactive Method

Several methods have been developed in reaction to some of the problems associated with earlier approaches. The first of these to be discussed is the visual interactive method developed by Korhonen and Laakso (1986). The method may be thought of as a blend of the Wierzbicki, the Zionts-Wallenius, and the Geoffrion, Dyer, and Feinberg methods, tied together by computer graphics on a microcomputer. It works as follows: (I have taken a bit of poetic license in describing the method).

1. Have the decision maker choose a desired level of each objective that he would like to obtain. This is like goal programming or Wierzbicki's approach. Then using a variation of the Wierzbicki approach, project this solution onto the efficient frontier, and designate the efficient solution as the incumbent solution.

2. Present the incumbent solution to the decision maker, and ask him to specify a new set of desired levels of each objective function that he would now like to achieve. Call this the desired solution, and construct a vector from the incumbent solution to the desired solution (in objective function space). Let that vector have infinite length, thereby extending through and beyond the desired solution.

3. Analogous to step one, project the vector constructed in step two onto the efficient frontier. The projection constitutes a piecewise-linear function along the efficient frontier. In terms of each objective, as we move along the projection,

accordingly, the objectives change in a piecewise-linear manner. Use simple computer graphics to show the changes that occur. See Figure 10.

4. Have the user do a line search (using the computer) along the projection to find his most preferred solution along the projection. As the user moves the cursor along the piecewise-linear segments, the screen displays the values for all of the objective functions. Designate the solution found as the incumbent solution, and go to step 2. If the incumbent solution remains the same, stop. An optimal solution has been found.

The computer implementation is on an IBM-PC and involves the representation of the objectives as piecewise continuous straight line functions. A different color is used for each objective. Some experimentation with the method is described by the authors,

Figure 10

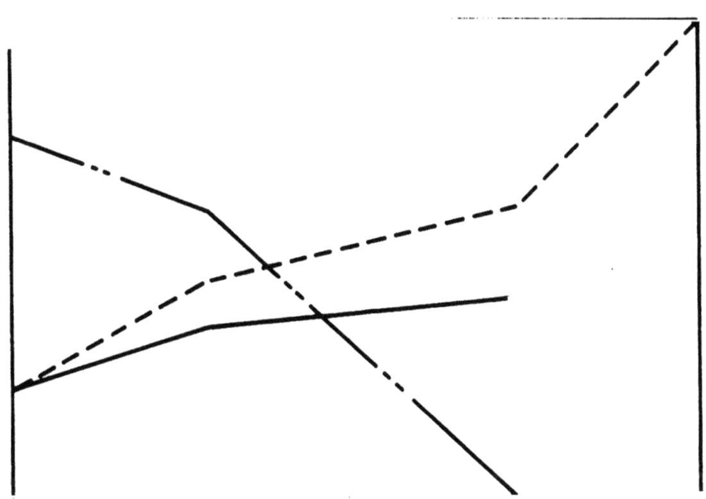

An Example of the Display in the Visual Interactive Method

and the empirical results with the method to date appear to be reasonable. The method is particularly attractive because of its implementation on a popular microcomputer, the IBM PC.

7. A Pareto Race

Based on the above approach, Korhonen and Wallenius (1987) have developed an approach that they call a Pareto Race. It continues in the tradition of Korhonen and Laakso, and is almost a video game. It combines the above ideas with the ideas of exploring all nondominated solutions, in the sense that the method explores a subset of nondominated solutions, as directed by the user. It may be used to solve a linear programming problem and involves having the decision maker, in a rough sense, explore the efficient frontier by "driving" around on it, thus the similarity to a video game.

In the case of two objectives, the corresponding efficient frontier may be thought of as one dimensional, and in the case of three objectives, the corresponding efficient frontier may be thought of as two dimensional, but for four or more objectives, the efficient frontier may be thought of as three dimensional or higher. (It is always one dimension less than the number of objectives.) Accordingly, except for two or three objective problems, it is not possible to represent the efficient frontier in a manner that may be visually represented on a plane. (The representation of a three dimensional problem on a plane may be thought of in terms of the weight space illustrated earlier.) Accordingly, the Pareto race uses bar graphs to represent the value of each objective. As the user moves around on the efficient frontier, the bar graphs (in color) change.

Using the analogy of "driving" around the frontier, the approach has certain functions that perform the movement along the frontier. These functions provide certain controls for the user (I use some poetic license in describing these functions; the authors describe these functions somewhat differently):

1. Provide movement in a preset direction. This corresponds to a unit movement along a projection described in the visual interactive method.

2. Increase or decrease the speed. This involves changing the stepsize of a unit movement in step one.

3. Change the minimum level of an objective. It may be set to a given value, or allowed to be free.

4. Change the direction of the path along the efficient frontier. Increase the component of a particular objective.

As with the visual interactive method, the Pareto Race has been implemented on an IBM PC/1 microcomputer. It is easy to use, and has been well received. The authors provide an illustration to a problem and several other applications.

Although the idea of the method and the way in which it is possible to move around the efficient frontier are interesting and worthwhile, the value of the method is greatly enhanced by the computer implementation and the graphics used.

8. A Discrete Alternatives Method

A method that has proved successful for solving the discrete alternatives problem is one developed by Korhonen, Wallenius and Zionts (1984) and Koksalan, Karwan, and Zionts (1984). See also Chung (1986). What we present here is an abbreviated version of the first reference. The authors assume a single decision maker who has an implicit quasi-concave increasing utility function of objectives.

The idea is to consider the convex hull of the solution points, and the convex set that it represents. By reference to Figure 1 and the solution points contained therein, observe that the extreme points of the convex hull of the solutions (including the origin O) are solutions O, A, D, H, J, and K. As stated earlier, though solutions B and F are nondominated, they are dominated by convex combinations of solutions (B by A and D, F by D and H).

The method proceeds as follows:

1. Choose an initial set of weights for the objectives. (In the absence of anything better, having scaled the solutions so that the maximum and minimum obtainable for each objective is one and zero, respectively, choose the weights for each objective the same.)

2. Identify the solution that maximizes the weighted sum of objectives (by enumeration) and designate the maximizing solution as the incumbent solution.

3. Using the incumbent solution as a reference solution, identify all of the adjacent efficient solutions to the incumbent solution. Ask the decision maker to compare the incumbent solution with an adjacent efficient solution. Whichever is less preferred is eliminated, and a convex cone is constructed as a result of the choice. Eliminate any solutions that are dominated by the new convex cone. (See below for an explanation of convex cones.)

4. When an adjacent efficient solution is preferred to the

incumbent solution, the adjacent solution becomes the incumbent. Go to step 3. When no adjacent efficient solutions are preferred to the incumbent, the incumbent solution is optimal.

The procedure eliminates solutions from consideration as a result of comparisons (any solution less preferred than another solution may be eliminated), and as a result of cone dominance. What cone dominance does is to construct a cone as a consequence of one solution being preferred to another. Let the preferred solution be solution A, and the less preferred solution be solution B. (More generally, there may be several solutions all preferred to another.) The authors prove that solution B is preferred to any solution in the cone (or halfline) emanating from B in the direction away from A, and consequently to any solutions dominated by the cone. See Figure 11.

Figure 11

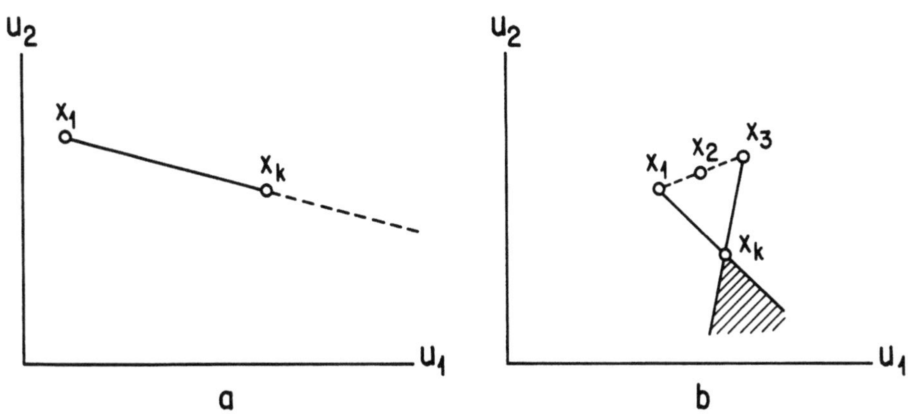

Some Illustrations of Cone Dominance

The method uses a linear proxy utility function (the weights) to generate solutions. This does not mean that the underlying utility function must be linear. As a result of solutions being eliminated, solutions that are convex dominated may be most preferred.

Chung describes an application of a derivative of this approach to a home purchase problem. He also has developed general proxy utility functions, and has the method choose the proxy utility function at each stage as a function of its apparent fit.

Conclusion

This presentation was designed as a brief introduction to the multiple criteria decision problem, with emphasis on the multiple objective linear programming problem. It includes a selection of methods from the plethora of those developed and presented in the literature. Our treatment was of necessity brief. I have included the methods I believe are important as developments or as stepping stones in the development of methods. For further information on what was presented, refer to the references. For additional information on these methods and new developments in the field, refer to the various journals in management science and decision support systems.

I have been working actively in the area of multiple criteria decision making over twenty years, with a short respite after my earliest work. I can honestly say that, from my perspective, the challenges in the field today are far greater than they have ever been. The computer technology is enabling us to come up with such approaches as numbers 6 and 7. I am sure that we will come up with even more powerful approaches. And the problems that we study become even more and more difficult. We are currently studying the group decision problem, because of its importance and difficulty, and its relationship with other important problems.

The critical issue regarding all of the methods of multiple criteria decision making is in the eating, so to speak, or in the application. The literature is replete with methods that have been developed and published, but have never been used to solve problems. Lest we think that is a blemish, we should reserve judgment until an appropriate later date. We shall see that these methods and methods yet to be developed are used and will be used.

REFERENCES

1. Benayoun, R., de Montgolfier, J., Tergny, J., and Larichev, O., "Linear Programming with Multiple Objective Functions: Step Method (STEM)," Mathematical Programming, 1, 1971, 615.

2. Charnes, A. and Cooper, W. W., Management Models and Industrial Applications of Linear Programming, John Wiley and Sons, New York, 1961.

3. Charnes, A. and Cooper, W. W., "Goal Programming and Multiple Objective Optimization - Part 1," European Journal of Operations Research, 1, 1977, 39.

4. Chung, H. W., Investigation of Discrete Multiple Criteria Decision Making and an Application to Home Buying, Unpublished Doctoral Dissertation, School of Management, State University of New York, Buffalo, 1986.

5. Deshpande, D., Investigations in Multiple Objective Linear Programming-Theory and an Application, Unpublished Doctoral Dissertation, School of Management, State University of New York at Buffalo, 1980.

6. Dyer, J., "A Time-Sharing Computer Program for the Solution of the Multiple Criteria Problem," Management Science, 19, 1973, 349.

7. Evans, J. P. and Steuer, R. E., "Generating Efficient Extreme Points in Linear Multiple Objective Programming: Two Algorithms and Computing Experience," in Cochrane and Zeleny, Multiple Criteria Decision Making, University of South Carolina Press, 1973.

8. Geoffrion, A. M., Dyer, J. S. and Feinberg, A., "An Interactive Approach for Multicriterion Optimization with an Application to the Operation of an Academic Department," Management Science, 19, 1972, 357.

9. Haimes, Y. Y., and Hall, W. A., "Multiobjectives in Water Resources Systems Analysis: The Surrogate Worth Trade Off Method," Water Resources Research, 10, 1974, 615.

10. Ijiri, Y., Management Goals and Accounting for Control, North-Holland Publishing Co., Amsterdam, and Rand McNally, Chicago, 1965.

11. Kallio, M., Lewandowski, A., and Orchard-Hays, W., "An Implementation of the Reference Point Approach for Multiobjective Optimization," Working Paper No. 80-35, International Institute for Applied Systems Analysis, Laxemburg, Austria, 1980.

12. Karwan, M. H. Zionts, S., Villareal, B., and R. Ramesh, "An Improved Interactive Multicriteia Integer Programming Algorithm," in Haimes, Y. Y. and Chan Kong, V., Decision Making with Multiple Objectives, Proceedings, Cleveland, Ohio, 1984, Lecture Notes in Economics and Mathematical Systems, Vol. 242, Springer-Verlag, (Berlin), 1985, pp. 261-271.

13. Keeney, R. L. and Raiffa, H., Decisions with Multiple Objectives Preferences and Value Tradeoffs, John Wiley and Sons, New York, 1976.

14. Koksalan, M., Karwan, M. H., and Zionts, S., "An Improved method for Solving Multicriteria Problems Involving Discrete Alternatives," IEEE Transactions on Systems, Man, and Cybernetics, Vol. 14, No. 1, January 1984, 24-34.

15. Korhonen, P. J., and Laakso, J., "A Visual Interactive Method for Solving the Multiple Criteria Problem," European Journal of Operational Research, 24, 1986, 277-287.

16. Korhonen, P. J., and Wallenius, J., A Pareto Race, Unpublished Paper, Helsinki School of Economics, 1987.

17. Korhonen, P., Wallenius, J., and Zionts, S., "Solving the Discrete Multiple Criteria Problem Using Convex Cones," Management Science, 30, 11, 1984, 1336-1345.

18. Lee, S. M., Goal Programming for Decision Analysis, Auerbach, Philadelphia, 1972.

19. Manheim, M. L., and Hall, F., "Abstract Representation of Goals: A Method for Making Decisions in Complex Problems," in Transportation: A Service, Proceedings of the Sesquicentennial Forum, New York Academy of Sciences American Society of Mechanical Engineers, New York 1967.

20. Miller, G., "The Magical Number Seven Plus or Minus Two: Some Limits on Our Capacity For Processing Information," in Psychological Review, 63, 1956, 81.

21. Ramesh, R., Multicriteria Integer Programming, Unpublished Doctoral Dissertation, Department of Industrial Engineering, State University of New York at Buffalo, 1985.

22. Roy, B., "Partial Preference Analysis and Decision Aid: The Fuzzy Criterion Concept," in Bell, D. E., Keeney, R. L. and Raiffa, H., eds., Conflicting Objectives in Decisions, International Series on Applied Systems Analysis, John Wiley and Sons, 1977, 442.

23. Steuer, R. E., "Multiple Objective Linear Programming with Interval Criterion Weights," Management Science, 23, 1977, 305.

24. Steuer, R. E., Multiple Criteria Optimization Theory, Computation, and Application, John Wiley and Sons, New York, 1986

25. Steuer, R. E., and , A. T., An Interactive Multiple Objective Linear Programming Approach to a Problem in Forest Management, Working Paper No. BA2, College of Business and Economics, University of Kentucky, 1976.

26. Steuer, R. E., and Wallace, M. J., Jr., "An Interactive Multiple Objective Wage and Salary Administration Procedure," in Lee, S. M., and Thorp, C. D., Jr., Eds., Personnel Management A Computer-Based System, Petrocelli, New York, 1978, 159.

27. Villareal, B., Multicriteria Integer Linear Programming, Doctoral Dissertation, Department of Industrial Engineering, State University of New York at Buffalo, 1979.

28. Wallenius, H., Wallenius, J., and Vartia, P., "An Approach to Solving Multiple Criteria Macroeconomic Policy Problems and an Application," Management Science, 24, 1978, 1021.

29. Wierzbicki, A. P., "The Use of Reference Objectives in Multiobjective Optimization," in G. Fandel and T. Gal (Eds.), Multiple Criteria Decision Making Theory and Application, Springer-Verlag, New York, 1980.

30. Yu, P. L. and Zeleny, M., "The Set of All Nondominated Solutions in the Linear Cases and a Multicriteria Simplex Method," Journal of Mathematical Analysis and Applications, 49, 1975, 430.

31. Zionts, S., "Integer Linear Programming with Multiple Objectives",Annals of Discrete Mathematics, 1, 1977, 551.

32. Zionts, S. and Wallenius, J., "An Interactive Programming Method for Solving the Multiple Criteria Problem," Management Science, 22, 1976, 652.

33. Zionts, S. and Wallenius, J., "An Interactive Multiple Objective Linear Programming Method for a Class of Underlying Nonlinear Utility Functions," Management Science, 29, 1983, 519.

THE EFFECTS OF RISK ON DECISION MAKING

James S. Dyer
Foster Parker Centennial Professor
College of Business Administration
University of Texas at Austin 78712/USA

1. Introduction

A fundamental assumption of classical decision theory is that an individual's attitude toward the risks involved in choices among uncertain alternatives can be captured by the shape of his utility function u(x) (e.g., see Pratt (1964)). A measurable value function v(x) can also be defined for this same individual over the same criterion to measure his strength of preference for incremental changes in the level of the criterion. The shape of this measurable value function can be interpreted as revealing the marginal value of additional units of the criterion. In general, there is no compelling reason to assume that u(x) = v(x) for the same decision maker, as discussed by Sarin (1982). If u(x) and v(x) are not the same, we argue that exploring the relationship between them may be a fruitful means of gaining insights regarding how risks effect the preferences, and ultimately the choice behavior, of individuals.

The modern concept of a utility function u(x) was formalized by von Neumann and Morgenstern (1947) for the case where the probability distribution over outcomes is given to the decision maker, and later generalized by Savage (1954) to the case where this probability distribution is determined subjectively by the decision maker. A utility function can be assessed by arbitrarily setting the utility values of the best and worst values of the criterion to be $u(x_1) = 1.0$ and $u(x_0) = 0.0$, respectively. The decision maker is then asked to specify the criterion value $x_{.5}$ so that he is indifferent between receiving $x_{.5}$ for sure and receiving a lottery with a

NATO ASI Series, Vol. F56
Multiple Criteria Decision Making and Risk Analysis
Using Microcomputers. Edited by B. Karpak and S. Zionts

half chance of obtaining x_1 and a half chance of receiving x_0. The utility of $x_{.5}$ is equal to the expected utility of the lottery, 0.5. In a similar manner, the values of $x_{.25}$ and $x_{.75}$ can be determined by reference to equiprobability lotteries between x_0 and $x_{.5}$, and between $x_{.5}$ and x_1. When a sufficient number of such points have been assessed from the decision maker, a curve can be faired through them to determine u(x) over the interval $[x_0,x_1]$. For additional details and some alternate utility function assessment procedures, see Farquhar (1984).

A measurable value function v(x) represents the differences in the decision maker's strength of preference between attribute levels. Suppose for levels of an attribute x_i, x_j, x_k, and x_l, we have $v(x_i) - v(x_j) = v(x_k) - v(x_l)$. Then the decision maker perceives the same difference in his strength of preference between x_i and x_j as between x_k and x_l. Axioms for the existence of a measurable value function are provided in Krantz, et al. (1971), and discussed by Dyer and Sarin (1979). One approach to assessing v(x) is by directly rating attribute levels on an arbitrary numerical scale, say [0, 100]. Other assessment approaches are discussed by Farquhar and Keller (1987). While the assessment of the utility function u(x) requires the explicit introduction of uncertain lotteries, the assessment of the measurable value function requires only deterministic assessment procedures.

In the next section, we review previous research related to the effort to characterize the effects of risk on the decisions made by an individual. Section 3 provides a summary of a more general theory that provides a basis for synthesizing much of the recent work on risky decision making. Also, the concept of risk is discussed in the context of this general model. Some implications of this extended model of risky decision making are described in Section 4. Conclusions and a brief summary are in Section 5.

2. Relative Risk Aversion

The concept of a risk attitude was first given a rigorous interpretation and quantified by Arrow (1971) and Pratt (1964). These authors compared an individual's von Neumann and Morgenstern utility function u(x) to actuarial behavior to derive the measure of a risk attitude. Simply stated, they define an individual as risk averse if he prefers the expected consequence of a gamble to the gamble itself, risk prone if he prefers the gamble to the expected consequence, and risk neutral if he is indifferent between the two. The Arrow-Pratt measure of a risk attitude for choices involving a consequence x is defined as

$$r(x) = -u''(x)/u'(x),$$

where u' is the first derivative and u" is the second derivative of the utility function.

If an individual's preferences are monotonically increasing in x, then $r(x) > 0$ for the risk averse individual, $r(x) < 0$ for the risk prone individual, and $r(x) = 0$ for the risk neutral individual. If the individual's preferences are monotonically decreasing in x, then these inequalities will be reversed. Since the traditional theory is presented based on the assumption of increasing preferences, we will adopt that convention here to avoid confusion.

One criticism of the use of the Arrow-Pratt theory as a measure of a risk attitude is that it simply compares preferences for lotteries to their actuarial values (or expected values) without consideration of the strength of preference that the individual feels for the consequences. Thus, this measure confounds his risk attitude with his strength of preference so that the actual effects of risk on his choices cannot be identified. In order to separate the effects of risk from those of strength of preference, Dyer and Sarin (1982) introduced the concept of relative risk aversion based on the comparison of the risky utility function u(x) with the strength of preference function v(x).

If u(x) and v(x) are both assessed over the same range for the same individual over the same interval of the criterion x, and $u(x) = v(x)$, then Dyer and Sarin reasoned that the introduction of uncertainty into the assessment procedure has no effect on the individual's preferences, and therefore the individual is relatively risk neutral. However, if $u(x) = v(x)$, then there exists some transformation function u_v defined on the range of v such that $u_v(v(x)) \neq u(x)$ for all values of the criterion x. They define the measure of the relative risk attitude r_v as

$$r_v = -u_v''(v(x))/u_v'(v(x))$$

and obtain the following result

Theorem (Dyer and Sarin (1982)): At $x \in X$, an individual is relatively risk averse if and only if $r_v > 0$, relatively risk prone if and only if $r_v < 0$, and relatively risk neutral if and only if $r_v = 0$.

An intuitive interpretation of this result is that if v(x) is concave, then u(x) will be even more concave if an individual is relatively risk averse. When this occurs, the utility function u_v defined on v(x) will also be concave. Some examples of relative risk attitudes are shown in Figure 1. Keller (1985) provides an empirical investigation of this concept that provides support for the notion that the preference representation function of an individual will often be different in risky and riskless cases. See also the empirical studies by Fischer (1977).

3. Modelling the Effects of Risk on Decisions

In the development of the concept of relative risk aversion, Dyer and Sarin implicitly assumed that an individual's utility function u(x) would differ from his strength of preference function v(x) because of the introduction of risk into the decision making task, but that u(x) would not change as a function of the level of the risk in the task. Here we consider a natural generalization of that model.

3.1. Preference as a Function of Risk

For simplicity, we refer to alternatives involving probabilistic outcomes as lotteries. Suppose we have a measure of risk R that can be used to order lotteries on the basis of their perceived level of riskiness. If the preferences of an individual for a lottery are influenced by both the riskiness of the lottery, measured by R, and the strength of preference that he feels for the outcomes, measured by v, then we can write his utility function u as

$$u_R(x) = f(v(x), R)$$

We refer to this representation of a utility function as the risk-value model.

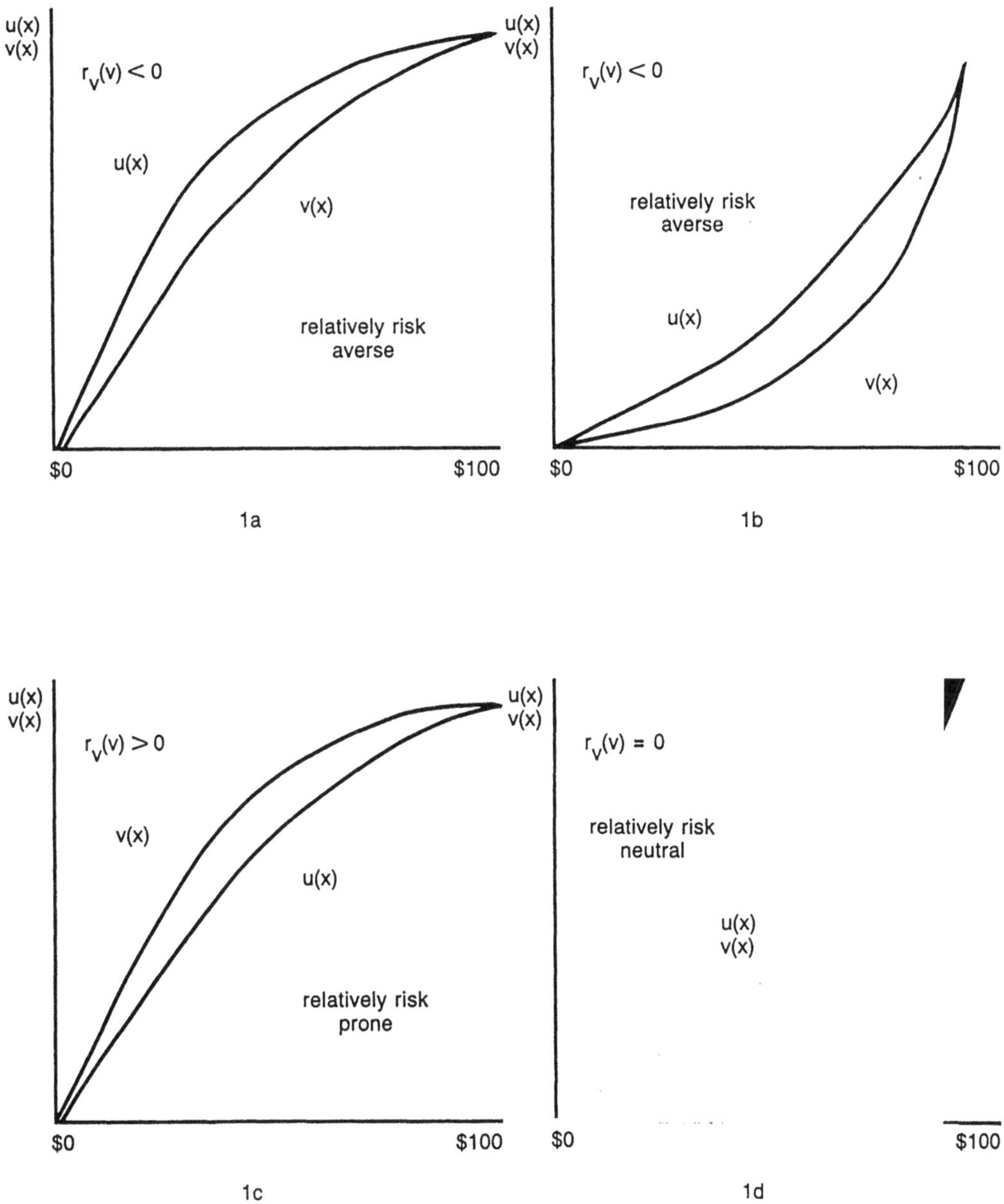

Figure 1. Examples of Relative Risk Aversion

The issues to be resolved include the actual form of f, which determines how v and R are to be combined, and the appropriate choice of R.

This extension of the concept of the utility function u to include a measure of the riskiness of the lottery as an explicit criterion may offer several interesting insights to the field of decision analysis, and may have practical implications as well. For example, new technologies are generally considered more "risky" because their outcomes cannot be predicted based on historical data, and because experts do not always agree in their predictions. Even if a new technology has a smaller expected number of lives lost, the variance associated with predictions of its outcomes will typically be much larger than the variance associated with the outcomes of an existing technology. If an individual uses a much more "risk averse" utility function u to evaluate alternatives that are perceived as more risky (in terms of the measure R), then this would help to explain public reluctance to accept new technologies even when individuals are convinced that the expected number of annual deaths that would result from the technology is low. This point is consistent with the finding of Slovic, et al. (1980) that individuals estimated the expected number of deaths that would result from accidents related to nuclear power to be low, but still rated that technology as very risky!

3.2 Measures of Risk

The concept of measuring risk presents some difficulties that can best be illustrated by reflecting on some simple examples that I have presented to several groups in lectures over the past few years. Consider each of the following in turn and respond to yourself regarding whether you consider the option to be a risky one:

1. Suppose you are offered $50 for sure. Is this risky?

2. Suppose you are offered an even chance lottery between $100 and $30. Is this risky?

3. Suppose you are offered an even chance lottery between -$100 and $30. Is this risky?

4. Suppose you are required to accept a $50 loss for sure. Is this risky?

5. Suppose you must choose between an even chance lottery between \$100 and \$30 and the alternative of \$50 for sure. Is this risky?

Individuals are unanimous saying that the offer of \$50 for sure is not risky, as we would expect since no probabilities are involved. The second option does introduce uncertainty, but most individuals argue that it is not risky because there is no chance of loss. The third option is unanimously considered to be risky by participants in this exercise, and is obtained by changing the sign of one of the outcomes in the previous lottery. The fourth option is not considered to be risky, even though it represents a loss. Finally, the fifth option is a choice between two alternatives, neither of which are considered risky by most respondents, but the <u>choice</u> is considered risky by most!

The objective of this simple thought experiment is to emphasize the point that the concept of "risk" may be a difficult one to measure. Notice that option 3 is considered to be a risky lottery because there exists a positive probability of a loss. However, option 5 is considered a risky decision, even though the alternatives themselves are not considered risky. Therefore, we can think of risk as the property of a lottery, but also think of risk as the property of a choice. Different definitions of risk may be necessary to account for these distinct notions.

Keller (1986) presents a review of twenty years of experiments that focus on the perception of risk by subjects. These experiments do demonstrate that subjects are able to consistently order lotteries when requested to rank them "in order of riskiness." These studies lend support to the effort to define a measure of riskiness R for lotteries that may depend on the amount of loss, amount of gain, probability distribution of losses, and probability distribution of gain.

As an example, Pollatsek and Tversky (1970) axiomatized a mean-variance model

$$R(X) = m(X) - wv(X)$$

In this model, R is the riskiness function for the random variable X, m is the mean, v is the variance, and w is a constant that reflects the trade-off between the mean and variance. Coombs and Bowen (1971) criticized the empirical validity of this model. Coombs then conducted a series of empirical studies (with several different researchers) from 1971 to 1981 in which he successively refined the models of riskiness. Coombs's approach was to propose a model and then proceed to test it empirically. His latest work is reported in Coombs and Lehner (1981a, 1981b).

Luce (1980) proposed several possible measures of the riskiness of a gamble. He used two key assumptions to derive the models. The first assumption says that the risk either increases additively or multiplicatively when a density function f(x) is multiplied by a constant. The second assumption stipulates the form of the risk function. The first form is analogous to the expected utility form and provides an expected risk function, and in the second form the density goes through a transformation before it is integrated to obtain R(x). These assumptions and the resulting four functional forms are summarized below;

	Form of the Risk Function R(f(x))	
	Transformation Used	
Risk Increases	Expected Risk	Density is Transformed
Additively	$\int \mathrm{Log}(x) f(x) dx$	$\int f(x) \log f(x) dx$
Multiplicatively	$\int x^{\theta} f(x) dx$	$\int f(x)^{1-\theta} dx$

Luce leaves the question of the reasonableness of these forms to empirical study.

Fishburn (1982,1984) provides a rigorous axiomatic development of a risk function. In one paper he considers only the loss domain as relevant for defining riskiness, and in the other he allows the possibility that gains could reduce the riskiness of a lottery. A lottery that only has gains will have no riskiness in Fishburn's models. Fishburn's most specialized model for the case in which only losses determine riskiness of a gamble is

$$\rho(\alpha, p) = \rho_1(\alpha) \int |x|^{\theta} dp(x), \theta > 0,$$

where α is the probability of loss, and p is the random variable of losses only. Thus, $\rho_1(\alpha)$ represents the riskiness due to the total loss probability α and the second term reflects how the amount of loss and the probability distribution for loss influence the risk.

We note that some of Fishburn's models are very similar to those of Luce. Luce specifies the exact functional form for each of his models, and thus these are easy to validate empirically. Fishburn's general models are difficult to validate directly, but the underlying axioms can be empirically tested. When Fishburn does make his models more specific they become similar to the multiplicative model of Luce.

4. The Risk-Value Model

The preceding discussion has provided a brief review of some of the literature regarding alternate proposals for the functional form of the measure of riskiness R in the risk value model $u = f(v,R)$. We now consider alternate choices for the function f, and explore some of the implications of this choice.

4.1 Risk-Value Model: The Additive Case

We begin with the additive case

$$u_l(x) = v(x) + wR_l$$

where the subscript l is used to denote that u_l is the utility function of an outcome x associated with a lottery l, and R_l is a measure of the riskiness of lottery l. This formulation makes explicit our assumption that the utility of the outcome x may differ depending on the lottery with which it is associated. In the special case where an individual's perception of the riskiness of a lottery is a constant for all risky lotteries, then u_l is a simple linear transformation of the strength of preference function v(x).

It is instructive to consider the three dimensional representation of this additive risk value model shown in Figure 2. Notice that the "shape" of the utility function stays the same as we move along the R axis, but that it is shifted up or down with respect to v(x). Therefore, at any given level of R, the Arrow-Pratt measure of the risk aversion of the decision maker with this representation will be identical.

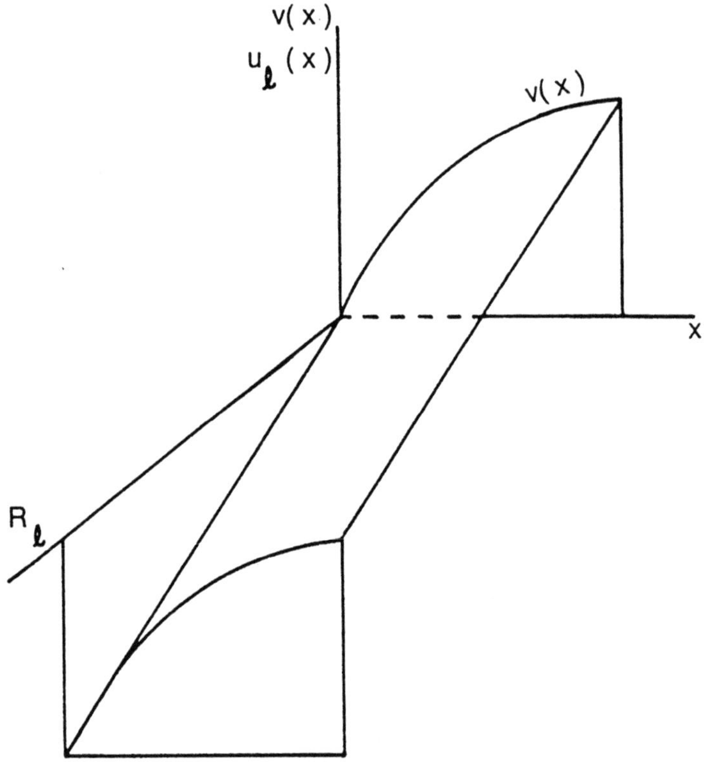

Figure 2. The Additive Risk-Value Model: $u_\ell(x) = v(x) + wR_\ell$

This general view of the additive risk-value model also allows the consideration of related work on generalizations of utility theory as special cases. For example, if we define

$$R_1 = E_1(v(x_i) - \bar{v})^2$$

where the expectation is taken over the outcome values x_i of the lottery, we obtain the mean-variance model of risky decision making proposed by Allais (1953). If we define

$$\begin{aligned} R_1 &= d(\bar{x} - x_i), \ \bar{x} > x_i \\ &= e(x_i - \bar{x}), \ x_i > \bar{x} \end{aligned}$$

we obtain the disappointment model by Bell (1985).

In both of these models, the risk is measured as a characteristic of a lottery. As we demonstrated in the fifth option used to illustrate the concept of risk, the choice among two alternatives may introduce additional perceptions of risk that are relevant to the decision maker. Suppose we consider a choice between two lotteries l1 and l2, and define the measure of risk

$$R_{l1,l2} = f(v(x) - v(y))$$

where x and y are outcomes obtained when an identical event occurs (e.g., "heads" occurs in a coin flip) in the lotteries l1 and l2 respectively. This is equivalent to the basic regret model of risky decision making proposed by Bell (1982). If l1 and l2 are independent lotteries, the definition

$$R_{l1,l2} = E_{Y/X}f(v(x) - v(y))$$

provides the appropriate generalization of Bell's regret model.

As this discussion suggests, several important models of risky decision making may be viewed as special cases of the linear risk-value model. In addition, this model has the potential for generating several other alternative models by introducing, for example, the models of risk proposed by Luce and by Fishburn and discussed in the previous section.

4.2. Risk-Value Model: The Multiplicative Case

Another important form of the risk-value model is the multiplicative case. In particular, consider the model

$$u_p = R(p)v(x)$$

where the measure of riskiness R is a function of the probability associated with receiving an outcome of x rather than being defined as a function of the entire probability distribution associated with the lottery. A graphical representation of this model is shown in Figure 3. Notice that the multiplicative form corresponds to a transformation of the scale of the utility function as the risk measure changes, but again the Arrow-Pratt measure of risk aversion for any level of R(p) will be the same.

The expression for expected utility obtained by substituting this model for the traditional von Neumann-Morgenstern utility function can be written

$$\sum_i p_i R(p_i) v(x_i)$$

where the subscript i denotes a particular state that occurs with probability p_i. Thus, the perceived risk associated with a p_i chance at receiving the outcome x_i influences preferences in a manner determined by $R(p_i)$. As a result, our preference for an outcome with a chance p_i of occurring will be different than our preference for the same outcome with a chance p_j of occurring, unless $R(p_i) = R(p_j)$, which will not be true in general.

As an alternative, suppose we group the terms in the expected value formulation in another way, and consider the interpretation of $\pi(p_i) = p_i R(p_i)$. Our choice of the notation π for this new function is not arbitrary, since with this alternate presentation of the model, the expected value formulation reduces to

$$\sum_i \pi(p_i) v(x_i)$$

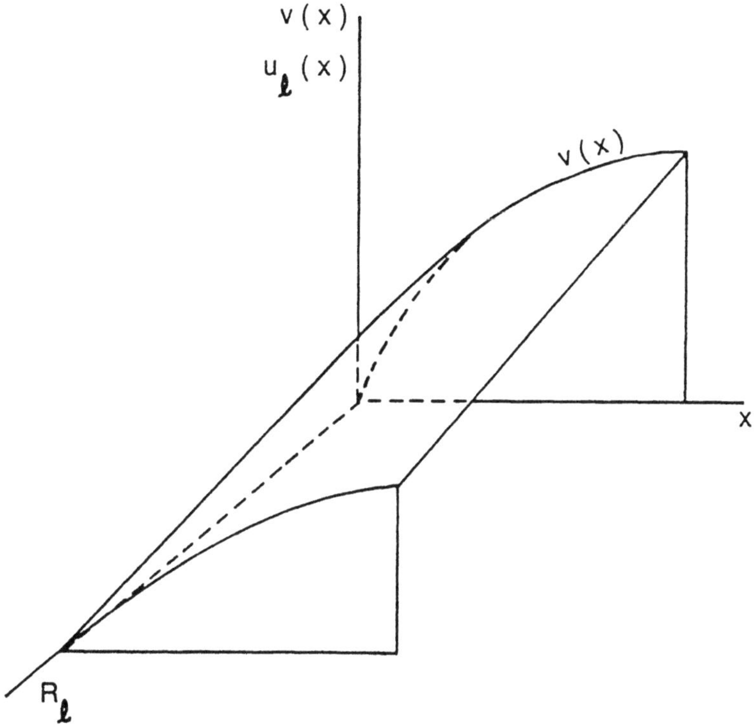

Figure 3. The Multiplicative Risk-Value Model : $u_p(x) = R(p)\,v(x)$

which is immediately recognizable as the prospect theory model of Kahneman and Tversky (1979).

This trivial manipulation of the model offers some insights that deserve comment. Kahneman and Tversky have motivated prospect theory on the premise that individuals subjectively weight probabilities when combining them with measures of value in order to determine the expected utility of a risky alternative. We offer an alternative explanation of their empirical results which leads to precisely the same mathematical formulation. That is, we suggest that individuals do not subjectively weight probabilities, but rather they alter their preferences as a function of these probabilities.

4.3 Risk-Value Theory: The Parameterized Case

As noted above, both the additive and multiplicative variations of the risk-value model share a characteristic that may be viewed as a shortcoming from a theoretical standpoint. Suppose the level of riskiness as measured by the function R is held constant in each of these models. It is easy to see in Figures 2 and 3 that both the additive form and the multiplicative form reduce to linear transformations of the strength of preference function v(x). Thus, the classical Arrow-Pratt measure of risk aversion $-u''/u'$ does not change in either model as a function of the riskiness of the alternative.

The riskiness of an alternative is added to v(x) as a reward or penalty term in the additive model, or appears as a scaling factor to reward or penalize an alternative in the multiplicative model. While both models have many interesting features and explain many of the common paradoxes of individual behavior, intuitively we would like to consider a model in which the measure $-u''/u'$ would change as a function of the level of perceived risk.

This observation inspired the research by Becker and Sarin (1987) on lottery dependent utility functions where the dependence of utility on the lottery is achieved by restricting the utility measure to a convenient parametric family of functions. The idea is to use each lottery to determine a specific parameter value, thus characterizing the utility function for each particular lottery. This form of utility function can be represented as

$$u_p(x) = u(x, c_p)$$

where c_p is a constant that depends on p. If c_p equals the same constant for all p, we obtain the classical von Neumann and Morgenstern utility theory as a special case. The value of c_p for p = 1.0 (certain outcomes) defines the measurable value function; e.g., $u(x,c_1) = v(x)$.

In order to provide a concrete example, Becker and Sarin assume that c_p is linear in probabilities. Thus, there exists a real function h(x) such that

$$c_p = \sum_x h(x)p(x) = E_p[h(x)]$$

They also provide examples of two functional forms for u(x,c), the exponential and the power. The former can be written

$$u(x,c_p) = \frac{1-e^{-c_p(x-x_0)/(x_1-x_0)}}{1-e^{c_p}} \qquad \text{if } c_p \neq 0$$

$$= \frac{x-x_0}{x_1-x_0} \qquad \text{if } c_p = 0$$

where x_1 and x_0 are the most and least preferred values of the criterion. A graphical representation of this function is provided in Figure 4, which illustrates the change in the "shape" of the utility function as the level of perceived riskiness changes.

5. Conclusions

Research on extended models of utility theory has been motivated by the empirical studies of actual decisions made by subjects confronted with risky alternatives. Many of these studies have demonstrated that subjects do not always make choices that are consistent with the axioms of classical utility theory. A number of authors have suggested specific modifications in models of decision making to represent this observed behavior.

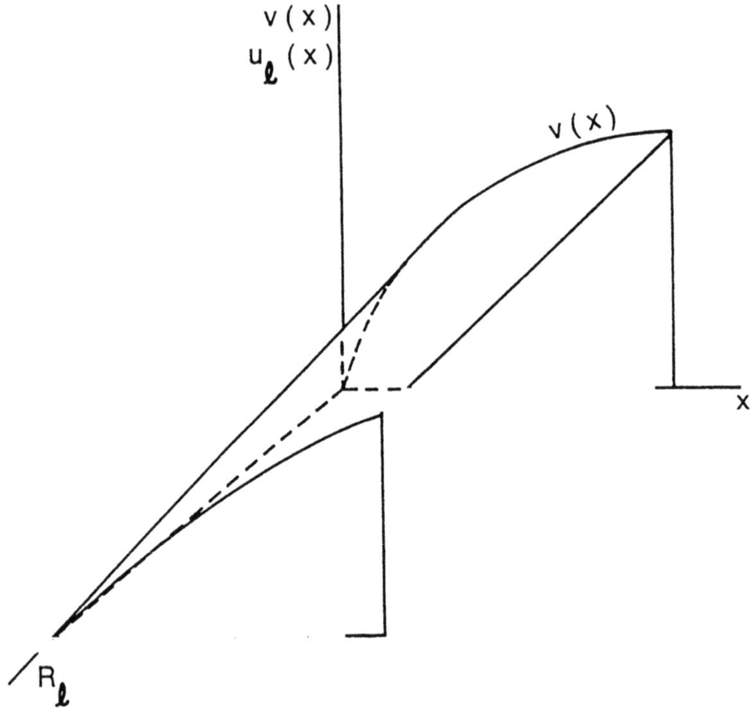

Figure 4. The Parametric Risk-Value Model : $u_p(x) = u(x, c_p)$

One objective of the development of the risk-value model is to provide a general theory that gives many of these specific models as special cases. This general theory allows a better understanding of the nature of the assumptions underlying the alternatives already suggested for modeling risky decision making, and may inspire the development of other models as well. As we have noted, an exploration of the variety of models of risk discussed in Section 3 coupled with various forms of the risk-value model could lead to interesting insights. The parametric model developed by Becker and Sarin is an example of a new contribution inspired by the implications of the general theory.

In addition, the assumptions that are implied by the additive, multiplicative, and parametric risk-value models could also be the basis for empirical tests. It would be interesting to discover which of these alternatives seems to provide a better empirical explanation for the actual choices made by subjects.

This discussion of the risk-value model is intended primarily as the basis for descriptive theories of risky decision making, rather than for prescriptive use. In general, we still feel that the axioms of von Neumann and Morgenstern are compelling as a basis for a guide to risky decisions. However, we acknowledge that in some special cases a decision maker may wish to explicitly consider the level of risk in an alternative in his decision making, and this more general theory could be applied as a guide to his decision making. This issue of the prescriptive versus the descriptive use of extensions to utility theory is discussed in more detail by Sarin (1988).

REFERENCES

Allais, M., "Le Comportement de l'homme Rationnel Devant le Risque: Critique des Postulats et Axiomes de l'ecole Americaine," *Econometrica*, *21*, 1953, pp. 503-546.

Arrow, K. J., *Essays in the Theory of Risk Bearing*, Markham, Chicago, Ill., 1971.

Becker, J. L., and R. K. Sarin, "Lottery Dependent Utility," *Management Science*, November, 1987, pp. 1367-1382.

Bell, D. E., "Regret in Decision Making Under Uncertainty," *Operations Research*, *30*, 1982, pp. 961-982.

__________, "Disappointment in Decision Making Under Uncertainty," *Operations Research*, *33*, January-February 1985, pp. 1-27.

Coombs, C. H., and J. N. Bowen, "A Test of the VE Theories of Risk and the Effect of the Central Limit Theorem," *Acta Psychologica*, *35*, 1971, pp. 15-28.

_______ and P. E. Lehner, "Evaluation of Two Alternative Models for a Theory of Risk: Part I," *Journal of Experimental Psychology, Human Perception, and Performance*, *7*, 1981a.

_______ and _______, "Evaluation of Two Alternative Models for a Theory of Risk II: The Bilinear Model and its Conjoint Analysis," University of Michigan, 1981.

Dyer, J. S., and R. K. Sarin, "Measurable Multiattribute Value Functions," *Operations Research*, *27*, 4, 1979.

_______ and _______, "Relative Risk Aversion," *Management Science*, *28*, 8, 1982

Farquhar, P. H., "Utility Assessment Methods," Management Science, 30, 11, November 1984, pp. 1283-1300.

_______ and R. Keller, "Preference Intensity Measurement," Decision Research Program, Technical Report 88-2, Graduate School of Industrial Administration, Carnegie Mellon University, Pittsburgh, Pennsylvania, 1988.

Fischer, G., "Convergent Validation of Decomposed Multi-attribute Utility Assessment Procedures for Risky and Riskless Decisions," Organizational Behavior and Human Performance, 18, 1977, pp. 295-315.

Fishburn, P. C., "Foundations of Risk Measurement II: Effects of Gains on Risk," Journal of Mathematical Psychology, 25, 1982.

________, Foundations of Risk Measurement I: Risk as Probable Loss," Management Science, 1984.

Kahnemen, D. and A. Tversky, "Prospect Theory: An Analysis of Decisions Under Risk," Econometrica, 47, 1979.

Keller, L. R., "An Empirical Investigation of Relative Risk Aversion," IEEE Transactions on Systems, Man, and Cybernetics, Vol. SMC-15, 4, July/August 1985, pp. 475-482.

_______, "Twenty Years of Experiments on the Riskiness of Gambles", Presentation at the Conference on Risk Measurement: Axiomatics, Experiments, and Applications, Lake Arrowhead, California, 1986.

Krantz, D. H., R. D. Luce, P. Suppes, and A. Tversky, Foundations of Measurement, Academic Press, New York, 1971.

Luce, R. D., "Several Possible Measures of Risk," Theory and Decision, 12, pp. 217-228, Correction, 13, p. 381, 1981.

Pollatsek, A. and A. Tversky, "A Theory of Risk," Journal of Mathematical Psychology," 1, 1970.

Pratt, J., "Risk Aversion in the Small and in the Large," Econometrica, 32, 1964, pp. 122-136.

Sarin, R. K., "Strength of Preference and Risky Choice," Operations Reasearch, 30, 5, 1982, pp. 982-997.

_______, "Analytical Issues in Decision Methodology," to appear in Decision and Organization Theory, Ira Horowitz (ed.), Kluwer-Nijhoff, 1988.

Savage, L. J., The Foundations of Statistics, John Wiley & Sons, New York, 1954.

Slovic, P., B. Fischhoff, and S. Lichtenstein, "Perceived Risk," in Societal Risk Assessment: How Safe is Safe Enough?," R. C. Schwing and W. A. Albers, Jr. (eds.), Plenum Press, New York, 1980.

von Neumann, J. and O. Morgenstern, Theory of Games and Economic Behavior, Princeton University Press, Princeton, N. J., 1947.

RATIONAL CHOICE AND THE FRAMING OF DECISIONS*

Amos Tversky
Department of Psychology
Stanford University
Stanford, California 94705, USA

Daniel Kahneman
Department of Psychology
University of California
Berkeley, California 94720, USA

Abstract

Alternative descriptions of a decision problem often give rise to different preferences, contrary to the principle of invariance that underlines the rational theory of choice. Violations of this theory are traced to the rules that govern the framing of decision and to the psychological principles of evaluation embodied in prospect theory. Invariance and dominance are obeyed when their application is transparent and often violated in other situations. Because these rules are normatively essential but descriptively invalid, no theory of choice can be both normatively adequate and descriptively accurate.

The modern theory of decision making under risk emerged from a logical analysis of games of chance rather than from a psychological analysis of risk and value. The theory was conceived as a normative model of an idealized decision maker, not as a description of the behavior of real people. In Schumpeter's words, it "has a much better claim to being called a logic of choice than a psychology of value" (1954, p. 1058).

The use of a normative analysis to predict and explain actual behavior is defended by several arguments. First, people

* This work has appeared in The Journal of Business, 1986, vol. 59(4), pt.2, pp. 251 - 278.

NATO ASI Series, Vol. F56
Multiple Criteria Decision Making and Risk Analysis
Using Microcomputers. Edited by B. Karpak and S. Zionts

are generally thought to be effective in pursuing their goals, particularly when they have incentives and opportunities to learn from experience. It seems reasonable, then, to describe choice as a maximization process. Second, competition favors rational individuals and organizations. Optimal decisions increase the chances of survival in a competitive environment, and a minority of rational individuals can sometimes impose rationality on the whole market. Third, the intuitive appeal of the axioms of rational choice makes it plausible that the theory derived from these axioms should provide an acceptable account of choice behavior.

The thesis of the present article is that, in spite of these a priori arguments, the logic of choice does not provide an adequate foundation for a descriptive theory of decision making. We argue that the deviations of actual behavior from the normative model are too widespread to be ignored, too systematic to be dismissed as random error, and too fundamental to be accommodated by relaxing the normative system. We first sketch an analysis of the foundations of the theory of rational choice and then show that the most basic rules of the theory are commonly violated by decision makers. We conclude from these findings that the normative and the descriptive analysis cannot be reconciled. A descriptive model of choice is presented, which accounts for preferences that are anomalous in the normative theory.

I. A Hierarchy of Normative Rules

The major achievements of the modern theory of decision under risk is the derivation of the expected utility rule from simple principles of rational choice that make no reference to long-run considerations (von Neumann and Morgenstern 1944). The axiomatic analysis of the foundations of expected utility theory reveals four substantive assumptions-cancellation, transitivity, dominance, and invariance-besides the more technical assumptions of comparability and continuity. The substantive assumptions can be ordered by their normative appeal, from the cancellation condition, which has been challenged by many theorists, to invariance, which has been accepted by all. We briefly discuss these assumptions.

Cancellation. The key qualitative property that gives rise to expected utility theory is the "cancellation" or elimination of any state of the world that yields the same outcome regardless of one's choice. This notion has been captured by different formal properties, such as the substitution axiom of von Neumann and Morgenstern (1944), the extended sure thing principle of Savage (1954), and the independence condition of Luce and Krantz (1971). Thus, if A is preferred to B, then the prospect of winning A if it rains tomorrow (and nothing otherwise) should be preferred to the prospect of winning B if it rains tomorrow because the two prospects yield the same outcome (nothing) if there is no rain tomorrow. Cancellation is necessary to represent preference between prospects as the maximization of expected utility. The main argument for cancellation is that only one state will actually be realized, which makes it reasonable to evaluate the outcomes of options separately for each state. The choice between options should therefore depend only on states in which they yield different outcomes.

Transitivity. A basic assumption in models of both risky and riskless choice is the transitivity of preference. This assumption is necessary and essentially sufficient for the representation of preference by an ordinal utility scale u such that A is preferred to B whenever $u(A) > u(B)$. Thus transitivity is satisfied if it is possible to assign to each option a value that does not depend on the other available options. Transitivity is likely to hold when the options are evaluated separately but not when the consequences of an option depend on the alternative to which it is compared, as implied, for example, by considerations of regret. A common argument for transitivity is that cyclic preferences can support a "money pump," in which the intransitive person is induced to pay for a series of exchanges that returns to the initial option.

Dominance. This is perhaps the most obvious principle of rational choice: if one option is better than another in one state and at least as good in all other states, the dominant option should be chosen. A slightly stronger condition--called stochastic dominance--asserts that, for unidimensional risky prospects, A is preferred to B if the cumulative distribution of A is to the right of the cumulative distribution of B. Dominance is both simpler and more compelling than cancellation and transitivity, and it serves as the cornerstone of the normative theory of choice.

Invariance. An essential condition for a theory of choice that claims normative status is the principle of invariance: different representations of the same choice problem should yield the same preference. That is, the preference between options should be independent of their description. Two characterizations that the decision maker, on reflection, would view as alternative descriptions of the same problem should lead to the same choice--even without the benefit of such reflection. This principle of invariance (or extensionality [Arrow 1982]), is so basic that it is tacitly assumed in the characterization of options rather than explicitly stated as a testable axiom. For, example, decision models that describe the objects of choice as random variables all assume that alternative representations of the same random variables should be treated alike. Invariance captures the normative intuition that variations of form that do not affect the actual outcomes should not affect the choice. A related concept, called consequentialism, has been discussed by Hammond (1985).

The four principles underlying expected utility theory can be ordered by their normative appeal. Invariance and dominance seem essential, transitivity could be questioned, and cancellation has been rejected by many authors. Indeed, the ingenious counterexamples of Allais (1953) and Ellsberg (1961) led several theorists to abandon cancellation and the expectation principle in favor of more general representations. Most of these models assume transitivity, dominance, and invariance (e.g., Hansson

1975; Allais 1979; Hagen 1979; Machina 1982; Quiggin 1982; Weber 1982; Chew 1983; Fishburn 1983; Schmeidler 1984; Segal 1984; Yaari 1984; Luce and Narens 1985). Other developments abandon transitivity but maintain invariance and dominance (e.g., Bell 1982; Fishburn 1982, 1984; Loomes and Sugden 1982). These theorists responded to observed violations of cancellation and transitivity by weakening the normative theory in order to retain its status as a descriptive model. However, this strategy cannot be extended to the failures of dominance and invariance that we shall document. Because invariance and dominance are normatively essential and descriptively invalid, a theory of rational decision cannot provide an adequate description of choice behavior.

We next illustrate failures of invariance and dominance and then review a descriptive analysis that traces these failures to the joint effects of the rules that govern the framing of prospects, the evaluation of outcomes, and the weighting of probabilities. Several phenomena of choice that support the present account are described.

II. Failures of Invariance

In this section we consider two illustrative examples in which the condition of invariance is violated and discuss some of the factors that produce these violations.

The first example comes from a study of preferences between medical treatments (McNeil et al. 1982). Respondents were given statistical information about the outcomes of two treatments of lung cancer.

The same statistics were presented to some respondents in terms of mortality rates and to others in terms of survival rates. The respondents then indicated their preferred treatment. The information was presented as follows.[1]

Problem 1(Survival frame)

Surgery: Of 100 people having surgery 90 live through the post operative period, 68 are alive at the end of the first year and 34 are alive at the end of five years.

Radiation Therapy: Of 100 people having radiation therapy all live through the treatment, 77 are alive at the end of the first year and 22 are alive at the end of five years.

Problem 1(Mortality frame)

Surgery: Of 100 people having surgery 10 die during surgery or the post-operative period, 32 die by the end of the first year and 66 die by the end of five years.

Radiation Therapy: Of 100 people having radiation therapy, none die during treatment, 23 die by the end of the first year and 78 die by the end of five years.

The inconsequential difference in formulation produced a marked effect. The overall percentage of respondents who favored radiation therapy rose from 18% in the survival frame (N = 247) to 44% in the mortality frame (N = 336). The advantage of radiation therapy over surgery evidently looms larger when stated as a reduction of the risk of immediate death from 10% to 0% rather than as an increase from 90% to 100% in the rate of survival. The framing effect was not smaller for experienced physicians or for statistically sophisticated business students

1. All problems are presented in the text exactly as they were presented to the participants in the experiments.

than for a group of clinic patients.

Our next example concerns decisions between conjunctions of risky prospects with monetary outcomes. Each respondent made two choices, one between favorable prospects and one between unfavorable prospects (Tversky and Kahneman 1981, p. 454). It was assumed that the two selected prospects would be played independently.

> Problem 2(N = 150). Imagine that you face the following pair of concurrent decisions. First examine both decisions, then indicate the options you prefer.
>
> Decisions (i) Choose between:
> A. a sure gain of $240 [84%]
> B. 25% chance to gain $1000 and 75% chance to gain nothing [16%]
>
> Decision (ii) Choose between:
> C. a sure loss $750 [13%]
> D. 75% chance to lose $1000 and 25% chance to lose nothing [87%]

The total number of respondents is denoted by N, and the percentage who chose each option is indicated in brackets. (Unless otherwise specified, the data were obtained from undergraduate students at Stanford University and at the University of British Columbia.) The majority choice in decision i is risk averse, while the majority choice in decision ii is risk seeking. This is a common pattern: choices involving gains are usually risk averse, and choices involving losses are often risk seeking--except when the probability of winning or losing is small (Fishburn and Kochenberger 1979; Kahneman and Tversky 1979; Hershey and Schoemaker 1980).

Because the subjects considered the two decisions

simultaneously, they expressed, in effect, a preference for the portfolio A and D over the portfolio B and C. However, the preferred portfolio is actually dominated by the rejected one! The combined options are as follows.

A & D: 25% chance to win $240 and 75% chance to lose $760.
B & C: 25% chance to win $250 and 75% chance to lose $750.

When the options are presented in this aggregated form, the dominant option is invariably chosen. In the format of problem 2, however, 73% of respondents chose the dominated combination A and D, and only 3% chose B and C. The contrast between the two formats illustrates a violation of invariance. The findings also support the general point that failures of invariance are likely to produce violations of stochastic dominance and vice versa.

The respondents evidently evaluated decisions i and ii separately in problem 2, where they exhibited the standard pattern of risk aversion in gains and risk seeking in losses. People who are given these problems are very surprised to learn that the combination of two preferences that they considered quite reasonable led them to select a dominated option. The same pattern of results was also observed in a scaled-down version of problem 2, with real monetary payoff (see Tversky and Kahneman 1981, p. 458).

As illustrated by the preceding examples, variations in the framing of decision problems produce systematic violations of invariance and dominance that cannot be defended on normative grounds. It is instructive to examine two mechanisms that could ensure the invariance of preferences: canonical representations and the use of expected actuarial value.

Invariance would hold if all formulations of the same prospect were transformed to a standard canonical representation (e.g., a cumulative probability distribution of the same random variable) because the various versions would then all be evaluated in the same manner. In problem 2, for example, invariance and dominance would both be preserved if the outcomes of the two decisions were aggregated prior to evaluation. Similarly, the same choice would be made in both versions of the medical problem if the outcomes were coded in terms of one dominant frame (e.g., rate of survival). The observed failures of invariance indicate that people do not spontaneously aggregate concurrent prospects or transform all outcomes into a common frame.

The failure to construct a canonical representation in decision problems contrasts with other cognitive tasks in which such representations are generated automatically and effortlessly. In particular, our visual experience consists largely of canonical representations: objects do not appear to change in size, shape, brightness, or color when we move around them or when illumination varies. A white circle seen from a sharp angle in dim light appears circular and white, not ellipsoid and grey. Canonical representations are also generated in the process of language comprehension, where listeners quickly recode much of what they hear into an abstract propositional form that no longer discriminates, for example, between the active and the the passive voice and often does not distinguish what was actually said from what was implied or presupposed (Clark and Clark 1977). Unfortunately, the mental machinery that transforms percepts and sentences into standard forms does not automatically apply to the process of choice.

Invariance could be satisfied even in the absence of a canonical representation if the evaluation of prospects were separately linear, or nearly linear, in probability and monetary value. If people ordered risky prospects by their actuarial values, invariance and dominance would always hold. In particular, there would be no difference between the morality and survival versions of the medical problem. Because the evaluation of outcomes and probabilities is generally non-linear, and because people do not spontaneously construct canonical representations of decisions, invariance commonly fails. Normative models of choice, which assume invariance, therefore cannot provide an adequate descriptive account of choice behavior. In the next section we present a descriptive account of risky choice, called prospect theory, and explore its consequences. Failures of invariance are explained by framing effects that control the representation of options, in conjunction with the nonlinearities of value and belief.

III. Framing and Evaluation of Outcomes

Prospect theory distinguishes two phases in the choice process: a phase of framing and editing, followed by a phase of evaluation (Kahneman and Tversky 1979). The first phase consists of a preliminary analysis of the decision problem, which frames the effective acts, contingencies and outcomes. Framing is controlled by the manner in which the choice problem is presented as well as by norms, habits, and expectancies of the decision maker. Additional operations that are performed prior to evaluation include cancellation of common components and the elimination of options that are seen to be dominated by others. In the second phase, the framed prospects are evaluated, and the prospect of highest value is selected. The theory distinguishes two ways of choosing between prospects: by detecting that one dominates another or by comparing their values.

For simplicity, we confine the discussions to simple gambles with numerical probabilities and monetary outcomes. Let $(x,p;y,q)$ denote a prospect that yields x with probability p and y with probability q and that preserves the status quo with probability $(1 - p - q)$. According to prospect theory, there are values $v(.)$, defined on gains and losses, and decisions weights $\pi(.)$, defined on stated probabilities, such that the overall value of the prospect equals $\pi(p)v(x) + \pi(q)v(y)$. A slight modification is required if all outcomes of a prospect have the same sign.[2]

The Value Function

Following Markowitz (1952), outcomes are expressed in prospect theory as positive or negative deviations (gains or losses) from a neutral reference outcome, which is assigned a value of zero. Unlike Markowitz, however, we propose that the value function is

---

2. If $p + q = 1$ and either $x > y > 0$ or $x < y < 0$, the value of a prospect is given by $v(y)+\pi(p)[v(x)-v(y)]$, so that decision weights are not applied to sure outcomes.

commonly S shaped, concave above the reference point, and convex below it, as illustrated in figure 1. Thus the difference in subjective value between a gain of $100 and a gain of $200 is greater than the subjective difference between a gain of $1100 and a gain of $1200. The same relation between value differences holds for the corresponding losses. The proposed function expresses the property that the effect of a marginal change decreases with the distance from the reference point in either direction. These hypotheses regarding the typical shape of the value function may not apply to ruinous losses or to circumstances in which particular amounts assume special significance.

A significant property of the value function, called *loss aversion*, is that the response to losses is more extreme than the response to gains. The common reluctance to accept a fair bet on the toss of a coin suggests that the displeasure of losing a sum of money exceeds the pleasure of winning the same amount. Thus the proposed value function is (i) defined on gains and losses, (ii) generally concave for gains and convex for losses, and, (iii) steeper for losses than for gains. These properties of the value function have been supported in many studies of risky choice involving monetary outcomes (Fishburn and Kochenberger 1979; Kahneman and Tversky 1979; Hershey and Schoemaker 1980; Payne, Laughhunn, and Crum 1980) and human lives (Tversky 1977; Eraker and Sox 1981; Tversky and Kahneman 1981; Fischhoff 1983). Loss aversion may also contribute to the observed discrepancies between the amount of money people are willing to pay for a good and the compensation they demand to give it up (Bishop and

Heberlein 1979; Knetsch and Sinden 1984). This effect is implied by the value function if the good is valued as a gain in the former context and as a loss in the latter.

Framing Outcomes

The framing of outcomes and the contrast between traditional theory and the present analysis are illustrated in the following problems.

> Problem 3 (N = 126): Assume yourself richer by $300 than you are today. You have to choose between a sure gain of $100 [72%]
> 50% chance to gain $200 and 50% chance to gain nothing [28%]
>
> Problem 4 (N =128): Assume yourself richer by $500 than you are today. You have to choose between
> a sure loss of $100 [36%]
> 50% chance to lose nothing and 50% to lose $200 [64%]

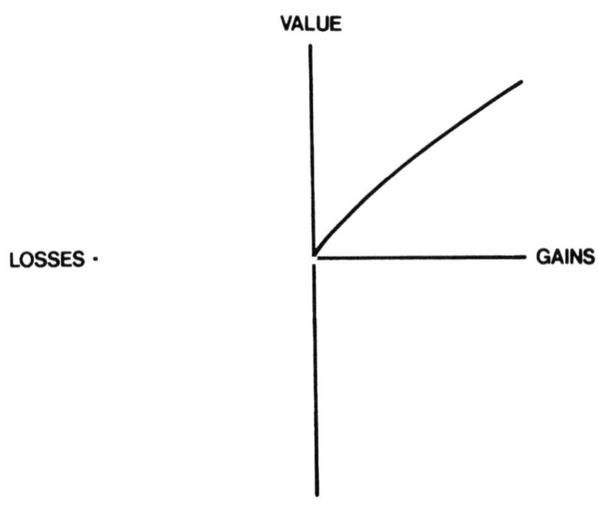

Fig. 1. A typical value function

As implied by the value function, the majority choice is risk averse in problem 3 and risk seeking in problem 4, although the two problems are essentially identical. In both cases one faces a choice between $400 for sure and even chance of $500 or $300. Problem 4 is obtained from problem 3 by increasing the initial endowment by $200 and subtracting this amount from both options. This variation has a substantial effect on preferences. Additional questions showed that variations of $200 in initial wealth have little or no effect on choices. Evidently, preferences are quite insensitive to small changes of wealth but highly sensitive to corresponding changes in reference point. These observations show that the effective carriers of values are gains and losses, or changes in wealth, rather than states of wealth as implied by the rational model.

The common pattern of preferences observed in problem 3 and 4 is of special interest because it violates not only expected utility theory but practically all other normatively based models of choice. In particular, these data are inconsistent with the model of regret advanced by Bell (1982) and by Loomes and Sugden (1982) and axiomatized by Fishburn 1982). This follows from the fact that problems 3 and 4 yield identical outcomes and an identical regret structure. Furthermore regret theory cannot accommodate the combination of risk aversion in problem 3 and risk seeking in problem 4--even without the corresponding changes in endowment that make the problems extensionally equivalent.

Shifts of reference can be induced by different decompositions of outcomes into risky and riskless components, as in the above problems. The reference point can also be shifted by a mere labeling of outcomes, as illustrated in the following problems (Tversky and Kahneman 1981, p. 453).

> Problem 5 (N = 152): Imagine that the U.S. is preparing for the outbreak of an unusual Asian disease, which is expected to kill 600 people. Two alternative programs to combat the disease have been proposed. Assume that the exact scientific estimates of the consequences of the programs are as follows:
>
> If Program A is adopted, 200 people will be saved. [72%]
>
> If Program B is adopted, there is 1/3 probability that 600 people will be saved, and 2/3 probability that no people will be saved. [28%]

In problem 5 the outcomes are stated in positive terms (lives saved), and the majority choice is accordingly risk averse. The prospect of certainly saving 200 lives is more attractive than a risky prospect of equal expected value. A second group of respondents was given the same cover story with the following descriptions of the alternative programs.

> Problem 6 (N =155):
> If Program C is adopted 400 people will die.[22%]
>
> If Program D is adopted there is 1/3 probability that nobody will die, and 2/3 probability that 600 people will die.[78%]

In problem 6 the outcomes are stated in negative terms (lives lost), and the majority choice is accordingly risk seeking. The certain death of 400 people is less acceptable than a two-thirds chance that 600 people will die. Problems 5 and 6, however, are essentially identical. They differ only in that the former is framed in terms of the number lives saved (relative to an expected loss of 600 lives if no action is taken), whereas the latter is framed in terms of the number of lives lost.

On several occasions we presented both versions to the same

respondents and discussed with them the inconsistent preferences evoked by the two frames. Many respondents expressed a wish to remain risk averse in the "lives saved" version and risk seeking in the "lives lost" version, although they also expressed a wish for their answers to be consistent. In the persistence of their appeal, framing affects resemble visual illusions more than computational errors.

Discounts and surcharges

Perhaps the most distinctive intellectual contribution of economic analysis is the systematic consideration of alternative opportunities. A basic principle of economic thinking is that opportunity costs and out-of-pocket costs should be treated alike. Preferences should depend only on relevant differences between options, not on how these differences are labeled. This principle runs counter to the psychological tendencies that make preferences susceptible to superficial variations in form. In particular, a difference that favors outcome A over outcome B can sometimes be framed either as an advantage of A or as a disadvantage of B by suggesting either B or A as the neutral reference point. Because of loss aversion, the difference will loom larger when A is neutral and B-A is evaluated as a loss than when B is neutral and A-B is evaluated as a gain. The significance of such variations of framing has been noted in several contexts.

Thaler (1980) drew attention to the effect of labeling a difference between two prices as a surcharge or a discount. It is easier to forgo a discount than to accept a surcharge because the same price difference is valued as a gain in the former case and as a loss in the latter. Indeed, the credit card lobby is said to insist that any price difference between cash and credit purchases should be labeled as cash discount rather than a credit surcharge. A similar idea could be invoked to explain why the price response to slack demand often takes the form of discounts or special concessions (Stigler and Kindahl 1970). Customers may be expected to show less resistance to the eventual cancellation of such temporary arrangements than to outright price increases. Judgments of fairness exhibit the same pattern (Kahneman, Knetsch, and Thaler, in this issue).

Schelling (1981) has described a striking framing effect in a context of tax policy. He points out that the tax table can be constructed by using as a default case either the childless family (as in fact done) or, say, the modal two-child family. The tax difference between a childless family and a two-child family is naturally framed as an exemption (for the two-child family) in the first frame and as a tax premium (on the childless family) in the second frame. This seemingly innocuous difference has a large effect on judgments of the desired relation between income, family size, and tax. Schelling reported that his students rejected the idea of granting the rich a larger exemption than the poor in the first frame but favored a larger tax premium on the childless rich than on the childless poor in the second frame. Because the exemption and the premium are alternative labels for the same tax difference in the two cases, the judgments violate invariance. Framing the consequences of a public policy in positive or in negative terms can greatly alter its appeal.

The notion of a money illusion is sometimes applied to workers' willingness to accept, in periods of high inflation, increases in nominal wages that do not protect their real income -although they would strenuously resist equivalent wage cuts in the absence of inflation. The essence of the illusion is that, whereas a cut in the nominal wage is always recognized as a loss, a nominal increase that does not preserve real income may be treated as a gain. Another manifestation of the money illusion was observed in a study of the perceived fairness of economic actions (Kahneman, Knetsch, and Thaler in press). Respondents in a telephone interview evaluated the fairness of the action described in the following vignette, which was presented in two versions that differed only in the bracketed clauses.

> A company is making a small profit. It is located in a community experiencing a recession with substantial unemployment [but no inflation/ and inflation of 12%]. The company decides to [decrease wages and salaries 7%/increase salaries only 5%] this year.

Although the loss of real income is very similar in the two versions, the proportion of respondents who judged the action of the company "unfair" or "very unfair" was 62% for a nominal reduction but only 22% for a nominal increase.

Bazerman (1983) has documented framing effects in experimental studies of bargaining . He compared the performance of experimental subjects when the outcomes of bargaining were formulated as gains or as losses. Subjects who bargained over the allocation of losses more often failed to reach an agreement and more often failed to discover a Pareto-optimal solution. Bazerman attributed these observations to the general propensity toward risk seeking in the domain of losses, which may increase the willingness of both participants to risk the negative consequences of a deadlock.

Loss aversion presents an obstacle to bargaining whenever the participants evaluate their own concessions as losses and the concessions obtained from the other party as gains. In negotiating over missiles, for example, the subjective loss of security associated with dismantling a missile may loom larger than the increment of security produced by a similar action on the adversary's part. If the two parties both assign a two-to-one ratio to the values of the concessions they make and of those they obtain, the resulting four-to-one gap may be difficult to bridge. Agreement will be much easier to achieve by negotiators who trade in "bargaining chips" that are valued equally, regardless of whose hand they are in. In this mode of trading, which may be common in routine purchases, loss aversion tends to disappear (Kahneman and Tversky 1984).

IV. The Framing and Weighting of Chance Events

In expected-utility theory, the utility of each possible outcome is weighted by its probability. In prospect theory the value of an uncertain outcome is multiplied by a decision weight $\pi(p)$, which is a monotonic function of p but is not a probability. The weighting function π has the following properties. First, impossible events are discarded, that is, $\pi(0) = 0$, and the scale is normalized so that $\pi(1) = 1$, but the function is not well behaved near the end points (Kahneman and Tversky 1979). Second, for low probabilities, $\pi(p) > p$, but $\pi(p) + \pi(1-p) \leq 1$ (subcertainty). Thus low probabilities are overweighted, moderate and high probabilities are underweighted, and the latter effect is more pronounced than the former. Third, $\pi(pr)/\pi(p) < \pi(pqr)/\pi(pq)$ for all $0 < p,q,r \leq 1$ (subproportionality). That is, for any fixed probability ratio r, the ratio of decision weights is closer to unity when the probabilities are low than when they are high, for example, $\pi(.1)/\pi(.2) > \pi(.4)/\pi(.8)$. A hypothetical weighting function that satisfies these properties is shown in figure 2. Its consequences are discussed in the next section.[3]

Nontransparent Dominance

The major characteristic of the weighting function is the overweighting of probability differences involving certainty and impossibility, for example, $\pi(1.0) - \pi(.9)$ or $\pi(.1) - \pi(0)$, relative to comparable differences in the middle of the scale,

3. The extension of the present analysis to prospects with many (nonzero) outcomes involves two additional steps. First, we assume that continuous (or multivalued) distributions are approximated, in the framing phase, by discrete distributions with a relatively small number of outcomes. For example, a uniform distribution on the interval (0, 90) may be represented by the discrete prospect (0, .1;10, .1;...;90, .1). Second, in the multiple-outcome case the weighting function, $\pi_p(p_i)$, must depend on the probability vector p, not only on the the component p_i, $i = 1,\ldots,n$. For example, Karmarkar (1978) uses the function $\pi_p(p_i) = \pi(p_i)/[\pi(p_1) + \ldots + \pi(p_n)]$. As in the two-outcome case, the weighting function is assumed to satisfy subcertainty, $\pi_p(p_1) + \ldots + \pi_p(p_n) \leq 1$, and subproportionality.

$\pi(.3) - \pi(.2)$. In particular, for small p, π is generally subadditive, for example, $\pi(.01) + \pi(.06) > \pi(.07)$. This property can lead to violations of dominance, as illustrated in the following pair of problems.

Problem 7 (N = 88). Consider the following two lotteries, described by the percentage of marbles of different colors in each box and the amount of money you win or lose depending on the color of a randomly drawn marble. Which lottery do you prefer?

Option A					
	90% white	6% red	1% green	1% blue	2% yellow
	$0	win $45	win $30	lose $15	lose $15
Option B					
	90% white	6% red	1% green	1% blue	2% yellow
	$0	win $45	win $45	lose $10	lose $15

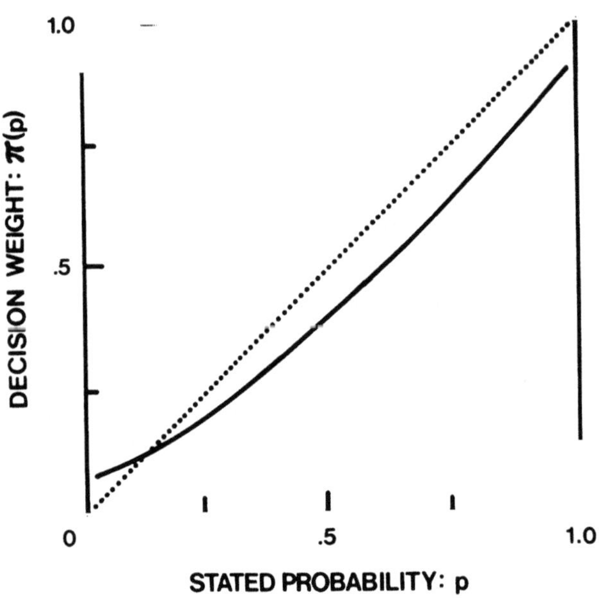

Fig.2. A typical weighting function

It is easy to see that option B dominates option A: for every color the outcome of B is at least as desirable as the outcome of A. Indeed, all respondents chose B over A. This observation is hardly surprising because the relation of dominance is highly transparent, so the dominated prospect is rejected without further processing. The next problem is effectively identical to problem 7, except that colors yielding identical outcomes (red and green in B and yellow and blue in A) are combined. We have proposed that this operation is commonly performed by the decision maker if no dominated prospect is detected.

Problem 8 (N = 124). Which lottery do you prefer?

Option C

90% white	6% red	1% green	3% yellow
$0	win $45	win $30	lose $15

Option D

90% white	7% red	1% green	2% yellow
$0	win $45	lose $10	lose $15

The formulation of problem 8 simplifies the options but masks the relation of dominance. Furthermore, it enhances the attractiveness of C, which has two positive outcomes and one negative, relative to D, which has two negative outcomes and one positive. As an inducement to consider the options carefully, participants were informed that one-tenth of them, selected at random, would actually play the gambles they chose. Although this announcement aroused much excitement, 58% of the participants chose the dominated alternative C. In answer to another question the majority of respondents also assigned a higher cash equivalent to C than to D. These results support the following propositions. (i) Two formulations of the same problem elicit different preferences, in violation of invariance. (ii) The dominance rule is obeyed when its application is transparent. (iii) Dominance is masked by a frame in which the inferior option yields more favorable outcome in an identified state of the world (e.g., drawing a green marble). (iv) The discrepant preferences are consistent with the subadditivity of decision weights. The role of transparency may be illuminated by a perceptual example. Figure 3 presents the well-known Müller-Lyer illusion: the top line appears longer than the bottom line, although it is in fact shorter. In figure 4, the same patterns are embedded in a rectangular frame, which makes it apparent that the protruding bottom line is longer than the top one. This judgment has the nature of an inference, in contrast to the perceptual impression that mediates judgment in figure 3. Similarly, the finer partition introduced in problem 7 makes it possible to conclude that option D is superior to C, without assessing their values. Whether the relation of dominance is detected depends on framing as well as on the sophistication and experience of the decision maker. The dominance relation in problems 8 and 1 could be transparent to a sophisticated decision maker, although it was not transparent to most of our respondents.

Certainty and Pseudocertainty

The over weighting of outcomes that are obtained with certainty relative to outcomes that are merely probable gives rise to violations of the expectation rule, as first noted by Allais

(1953). The next series of problems (Tversky and Kahneman 1981, p. 455) illustrates the phenomenon discovered by Allais and its relation to the weighting of probabilities and to the framing of chance events. Chance events were realized by drawing a single marble from a bag containing a specified number of favorable and unfavorable marbles. To encourage thoughtful answers, one-tenth of the participants, selected at random, were given an opportunity to play the gambles they chose. The same respondents answered problems 9-11, in that order.

> Problem 9 (N = 77). Which of the following options do you prefer?
> A. a sure gain of $30[78%]
> B. 80% chance to win $45 and 20% chance to win nothing [22%]
>
> Problem 10 (N = 81). Which of the following options do you prefer?
> C. 25% chance to win $30 and 75% chance to win nothing [42%]
> D. 20% chance to win $45 and 80% chance to win nothing [58%]

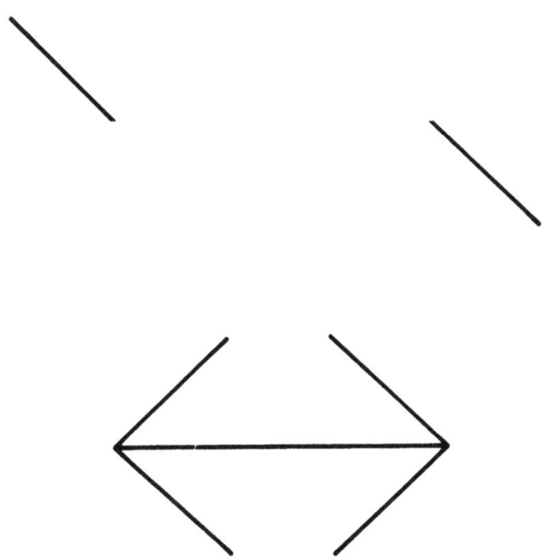

Fig. 3. The Müller-Lyer illusion

Note that problem 10 is obtained from problem 9 by reducing the probabilities of winning by a factor of four. In expected utility theory a preference for A over B in problem 9 implies a preference for C over D in problem 10. Contrary to this prediction, the majority preference switched from the lower prize ($30) to the higher one ($45) when the probabilities of winning were substantially reduced. We called this phenomenon the *certainty effect* because the reduction of the probability of winning from certainty to .25 has a greater effect than the corresponding

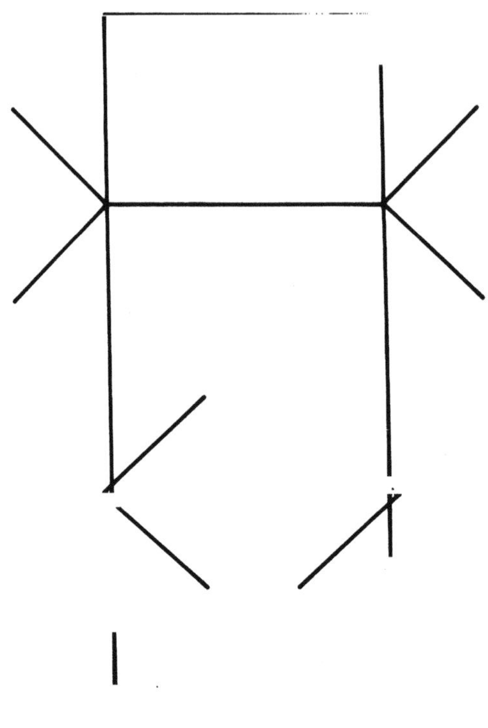

Fig. 4. A transparent version of the Müller-Lyer illusion

reduction from .8 to .2. In prospect theory, the modal choice in problem 9 implies $v(45)\pi(.80) < v(30)\pi(1.0)$, whereas the modal choice in problem 10 implies $v(45)\pi(.20) > v(30)\pi(.25)$. The observed violation of expected utility theory, then, is implied by the curvature of π (see fig. 2) if

$$\frac{\pi(.20)}{\pi(.25)} > \frac{v(30)}{v(45)} > \frac{\pi(.80)}{\pi(1.0)}$$

Allais's problem has attracted the attention of numerous theorists, who attempted to provide a normative rationale for the certainty effect by relaxing the cancellation rule (see, e.g., Allais 1979; Fishburn 1982, 1983; Machina 1982; Quiggin 1982; Chew 1983). The following problem illustrates a related phenomenon, called the *pseudocertainty effect*, that cannot be accommodated by relaxing cancellation because it also involves a violation of invariance.

> Problem 11 (N = 85): Consider the following two stage game. In the first stage, there is a 75% chance to end the game without winning anything, and a 25% chance to move into the second stage. If you reach the second stage you have a choice between:
>
> E. a sure win of $30[74%]
> F. 80% chance to win $45 and 20% chance to win nothing [26%]
>
> Your choice must be made before the outcome of the first stage is known.

Because there is one chance in four to move into the second stage, prospect E offers a .25 probability of winning $30, and prospect F offers a .25 x .80 = .20 probability of winning $45. Problem 11 is therefore identical to problem 10 in terms of probabilities and outcomes. However, the preferences in the two problems differ: most subjects made a risk-averse choice in problem 11 but not in problem 10. We call this phenomenon the pseudocertainty effect because an outcome that is actually uncertain is weighted as if it were certain. The framing of problem 11 as a two-stage game encourages respondents to apply cancellations: the event of failing to reach the second stage is discarded prior to evaluation because it yields the same outcomes in both options. In this framing, problems 11 and 9 are evaluated alike.

Although problems 10 and 11 are identical in terms of final outcomes and their probabilities, problem 11 has a greater potential for inducing regret. Consider a decision maker who chooses F in problem 11, reaches the second stage, but fails to win the prize. This individual knows that the choice of E would have yielded a gain of $30. In problem 10, on the other hand, an individual who chooses D and fails to win cannot know with certainty what the outcome of the other choice would have been. This difference could suggest an alternative interpretation of the pseudocertainty effect in terms of regret (e.g., Loomes and Sugden 1982). However, the certainty and the pseudocertainty effects were found to be equally strong in a modified version of problems 9-11 in which opportunities for regret were equated across problems. This finding does not imply that considerations of regret play no role in decisions. (For examples, see Kahneman and Tversky [1982, p.710].) It merely indicates that Allais's example and the pseudocertainty effect are primarily controlled by the nonlinearity of decision weights and the framing of contingencies rather than by the anticipation of regret.[4]

The certainty and pseudocertainty effects are not restricted to monetary outcomes. The following problem illustrates these phenomena in a medical context. The respondents were 72 physicians attending a meeting of the California Medical Association. Essentially the same pattern of responses was obtained from a larger group (N = 180) of college students.

> Problem 12 (N =72). In the treatments of tumors there is sometimes a choice between two types of therapies: (i)a radical treatment such as extensive surgery, which involves some risk of imminent death, (ii) a moderate treatment, such as limited surgery or radiation therapy. Each of the following problems describes the possible outcome of two alternative treatments, for three different cases. In considering each case suppose the patient is a 40-year-old male. Assume that without treatment death is imminent (within a month) and that only one of the treatments can be applied. Please indicate the treatment you would prefer in each case.
>
> Case 1
>
> Treatment A: 20% chance of imminent death and 80% chance of normal life, with an expected longevity of 30 years. [35%]
>
> Treatment B: certainty of a normal life, with an expected longevity of 18 years. [65%]
>
> Case2
>
> Treatment C: 80% chance of imminent death and 20% chance of normal life, with an expected longevity of 30 years.[68%]
>
> Treatment D: 75% chance of imminent death and 25% chance of normal life, with an expected longevity of 18 years.[32%]

---

4. In the modified version --problems 9'- 11'--the probabilities of winning were generated by drawing a number from a bag containing 100 sequentially numbered tickets. In problem 10', the event associated with winning $45 (drawing a number between one and 20) was included in the event associated with winning $30 (drawing a number between one and 25). The sequential setup of problem 11 was replaced by the simultaneous play of two chance devices: the roll of a die(whose outcome determines whether the game is on) and the drawing of a numbered ticket from a bag. The possibility of regret now exists in all three problems, and problem 10' and 11' no longer differ in this respect because a decision maker would always know the outcomes of alternative choices. Consequently, regret theory cannot explain either the certainty effect (9' vs. 10') or the pseudo-certainty effect (10' vs. 11') observed in the modified problems.

Case 3

Consider a new case where there is a 25% chance that the tumor is treatable and a 75% chance that it is not. If the tumor is not treatable, death is imminent. If the tumor is treatable, the outcomes of the treatment are as follows:

Treatment E: 20% chance of imminent death and 80% chance of normal life, with an expected longevity of 30 years.[32%]

Treatment F: certainty of normal life, with an expected longevity of 18 years. [68%]

The three cases of this problem correspond, respectively, to problems 9-11, and the same pattern of preferences is observed. In case 1, most respondents make a risk-averse choice in favor of certain survival with reduced longevity. In case 2, the moderate treatment no longer ensures survival, and most respondents choose the treatment that offers the higher expected longevity. In particular, 64% of the physicians who chose B in case 1 selected C in case 2. This is another example of Allais's certainty effect.

The comparison of cases 2 and 3 provides another illustration of pseudocertainty. The cases are identical in terms of the relevant outcomes and their probabilities, but the preferences differ. In particular, 56% of the physicians who chose C in case 2 selected F in case 3. The conditional framing induces people to disregard the event of the tumor not being treatable because the two treatments are equally ineffective in this case. In this frame, treatment F enjoys the advantage of pseudocertainty. It appears to ensure survival, but the assurance is conditional on the treatability of the tumor. In fact there is only a .25 chance of surviving a month if this option is chosen.

The conjunction of certainty and pseudocertainty effects has significant implications for the relation between normative and descriptive theories of choice. Our results indicate that cancellation is actually obeyed in choices -in those problems that make its application transparent. Specifically, we find that people make the same choices in problems 11 and 9 and cases 3 and 1 of problem 12. Evidently people "cancel" an event that yields the same outcomes for all options, in two-stage or nested structures. Note that in these examples cancellation is satisfied in problems that are formally equivalent to those in which it is violated. The empirical validity of cancellation therefore depends on the framing of the problems.

The present concept of framing originated from the analysis of Allais's problems by Savage (1954, pp.101-4) and Raiffa (1968, pp. 80-86), who reframed these examples in an attempt to make the application of cancellation more compelling. Savage and Raiffa were right: naive respondents indeed obey the cancellation axiom when its application is sufficiently transparent.[5] However, the contrasting preferences in different versions of the same choice (problem 10 and 11 and cases 2 and 3 of problem 12) indicate that people do not follow the same axiom when its application is not

---

5. It is noteworthy that the conditional framing used in problems 11 and 12 (case 3) is much more effective in eliminating the common responses to Allais's paradox than the partition framing introduced by Savage (see, e.g.,Slovic and Tversky 1974). This is probably due to the fact that the conditional framing makes it clear that the critical options are identical--after eliminating the state whose outcome does not depend on one's choice (i.e., reaching the second stage in problem 11, an untreatable tumor in problem 12, case 3).

transparent. Instead, they apply (non-linear) decision weights to the probabilities as stated. The status of cancellation is therefore similar to that of dominance: both rules are intuitively compelling as abstract principles of choice, consistently obeyed in transparent problems and frequently violated in nontransparent ones. Attempts to rationalize the preferences in Allais's example by discarding the cancellation axiom face a major difficulty: they do not distinguish transparent formulations in which cancellation is obeyed from nontransparent ones in which it is violated.

V. Discussion

In the preceding sections we challenged the descriptive validity of the major tenets of expected utility theory and outlined an alternative account of risky choice. In this section we discuss alternative theories and argue against the reconciliation of normative and descriptive analyses. Some objections of economists to our analysis and conclusions are addressed.

Descriptive and Normative Considerations

Many alternative models of risky choice, designed to explain the observed violations of expected utility theory, have been developed in the last decade. These models divide into the following four classes. (i) Nonlinear functionals (e.g., Allais 1953, 1979; Machina 1982) are obtained by eliminating the cancellation condition altogether. These models do not have axiomatizations leading to a (cardinal) measurement of utility, but they impose various restrictions (i.e., differentiability) on the utility functional. (ii) The expectations quotient model (axiomatized by Chew and MacCrimmon 1979; Weber 1982; Chew 1983; Fishburn 1983) replaces cancellation by a weaker substitution axiom and represents the value of prospect by the ratio of two linear functionals. (iii) Bilinear models with nonadditive probabilities (e.g., Kahneman and Tversky 1979; Quiggin 1982; Schmeidler 1984; Segal 1984; Yaari 1984; Luce and Narens 1985) assume various restricted versions of cancellation (or substitution) and construct a bilinear representation in which the utilities of outcomes are weighted by a nonadditive probability measure or by some nonlinear transform of the probability scale. (iv) Nontransitive models represent preferences by a bivariate utility function. Fishburn (1982,1984) axiomatized such models, while Bell (1982) and Loomes and Sugden (1982) interpreted them in terms of expected regret. For further theoretical developments, see Fishburn (1985)

The relation between models and data is summarized in table 1. The stub column lists the four major tenets of expected utility theory. Column 1 lists the major empirical violations of these tenets and cites a few representative references. Column 2 lists the subset of models discussed above that are consistent with the observed violations.

TABLE 1 Summary of Empirical Violations and Explanatory models

Tenet	Empirical Violation	Explanatory Model
Cancellation	Certainty effect (Allais 1953, 1979; Kahneman and Tversky 1979) (problems 9-10, and 12 [cases 1&2])	All models
Transitivity	Lexicographic semiorder(Tversky 1969) Preference reversals (Slovic and Lichtenstein 1983)	Bivariate models
Dominance	Contrasting risk attitudes(problem 2) Subadditive decision weights(problem 8)	Prospect theory
Invariance	Framing effects (Problems 1, 3-4, 5-6,7-8,10-11 and 12)	Prospect theory

The conclusions of table 1 may be summarized as follows. First, all the above models (as well as some others) are consistent with the violations of cancellation produced by the certainty effect.[6] Therefore, Allais's "paradox" cannot be used to compare or evaluate competing nonexpectation models. Second, bivariate (nontransitive) models are needed to explain observed intransitivities. Third, only prospect theory can accommodate the observed violations of (stochastic) dominance and invariance. Although some models (e.g., Loomes and Sugden 1982; Luce and Narens 1985) permit some limited failures of invariance, they do not account for the range of framing effects described in this article.

Because framing effects and the associated failures of invariance are ubiquitous, no adequate descriptive theory can ignore these phenomena. On the other hand, because invariance (or extensionality) is normatively indispensable, no adequate

6. Because the present article focuses on prospects with known probabilities, we do not discuss the important violations of cancellation due to ambiguity (Ellsberg 1961).

prescriptive theory should permit its violation. Consequently, the dream of constructing a theory that is acceptable both descriptively and normatively appears unrealizable (see also Tversky and Kahneman 1983).

Prospect theory differs from the other models mentioned above in being unabashedly descriptive and in making no normative claims. It is designed to explain preferences, whether or not they can be rationalized. Machina (1982, p. 292) claimed that prospect theory is "unacceptable as a descriptive model of behavior toward risk" because it implies violations of stochastic dominance. But since the violations of dominance predicted by the theory have actually been observed (see problems 2 and 8), Machina's objection appears invalid.

Perhaps the major finding of the present article is that the axioms of rational choice are generally satisfied in transparent situations and often violated in nontransparent ones. For example, when the relation of stochastic dominance is transparent (as in the aggregated version of problem 2 and in problem 7), practically everyone selects the dominant prospect. However, when these problems are framed so that the relation of dominance is no longer transparent (as in the segregated version of problem 2 and problem 8), most respondents violate dominance, as predicted. These results contradict all theories that imply stochastic dominance as well as others (e.g., Machina 1982) that predict the same choices in transparent and nontransparent contexts. The same conclusion applies to cancellation, as shown in the discussion of pseudocertainty. It appears that both cancellation and dominance have normative appeal, although neither one is descriptively valid.

The present results and analysis--particularly the role of transparency and the significance of framing--are consistent with the conception of bounded rationality originally presented by Herbert Simon (see, e.g., Simon 1955, 1978; March 1978; Nelson and Winter 1982). Indeed, prospect theory is an attempt to articulate some of the principles of perception and judgment that limit the rationality of choice.

The introduction of psychological considerations (e.g., framing) both enriches and complicates the analysis of choice. Because the framing of decisions depends on the language of presentation , on the context of choice, and on the nature of the display, our treatment of the process is necessarily informal and incomplete. We have identified several common rules of framing and we have demonstrated their effects on choice, but we have not provided a formal theory of framing. Furthermore, the present analysis does not account for all the observed failures of transitivity and invariance. Although some intransitivities (e.g., Tversky 1969) can be explained by discarding small differences in the framing phase, and others (e.g., Raiffa 1968, p. 75) arise from the combination of transparent and nontransparent comparisons, there are examples of cyclic preferences and context effects (see e.g., Slovic, Fischhoff, and Lichtenstein 1982; Slovic and Lichtenstein 1983) that require additional explanatory mechanisms (e.g., multiple reference points and variable weights). An adequate account of choice cannot ignore these effects of framing and context, even if they are normatively distasteful and mathematically intractable

Bolstering Assumptions

The assumption of rationality has a favored position in economics. It is accorded all the methodological privileges of a self-evident truth, a reasonable idealization, a tautology, and a null hypothesis. Each of these interpretations either puts the hypothesis of rational action beyond question or places the burden of proof squarely on any alternative analysis of belief and choice. The advantage of the rational model is compounded because no other theory of judgment and decision can ever match it in scope, power, and simplicity.

Furthermore, the assumption of rationality is protected by a formidable set of defenses in the form of bolstering assumptions that restrict the significance of any observed violation of the model. In particular, it is commonly assumed that substantial violations of the standard model are (i) restricted to insignificant choice problems, (ii) quickly eliminated by learning, or (iii) irrelevant to economics because of the corrective function of the market forces. Indeed, incentives sometimes improve the quality of decisions, experienced decision makers often do better than novices, and the forces of arbitrage and competition can nullify some effects of error and illusion. Whether these factors ensure rational choices in any particular situation is an empirical issue, to be settled by observation, not by supposition.

It has frequently been claimed (see, e.g., Smith 1985) that the observed failures of rational models are attributable to the cost of thinking and will thus be eliminated by proper incentives. Experimental findings provide little support for this view. Studies reported in the economic and psychological literature have shown that errors that are prevalent in responses to hypothetical questions persist even in the presence of significant monetary payoffs. In particular, elementary blunders of probabilistic reasoning (Grether 1980; Tversky and Kahneman 1983), major inconsistencies of choice (Grether and Plott 1979; Slovic and Lichtenstein 1983), and violations of stochastic dominance in nontransparent problems (see problem 2 above) are hardly reduced by incentives. The evidence that high stakes do not always improve decisions is not restricted to laboratory studies. Significant errors of judgment and choice can be documented in real world decisions that involve high stakes and serious deliberation. The high rate of failures of small businesses, for example, is not easily reconciled with the assumptions of rational expectations and risk aversion.

Incentives do not operate by magic: they work by focusing attention and by prolonging deliberation. Consequently, they are more likely to prevent errors that arise from insufficient attention and effort than errors that arise from misperception or faulty intuition. The example of visual illusion is instructive. There is no obvious mechanism by which the mere introduction of incentives (without the added opportunity to make measurements) would reduce the illusion observed in figure 3, and the illusion vanishes--even in the absence of incentives--when the display is altered in figure 4. The corrective power of incentives depends on the nature of the particular error and cannot be taken for granted.

The assumption of rationality of decision making is often defended by the argument that people will learn to make correct decisions and sometimes by the evolutionary argument that irrational decision makers will be driven out by the rational ones. There is no doubt that learning and selection do take place and tend to improve efficiency. As in the case of incentives, however, no magic is involved. Effective learning takes place only under certain conditions: it requires accurate and immediate feedback about the relation between the situational conditions and the appropriate response. The necessary feedback is often lacking for the decisions made by managers, entrepreneurs, and politicians because (i) outcomes are commonly delayed and not easily attributable to a particular action; (ii) variability in the environment degrades the reliability of the feedback, especially where outcomes of low probability are involved; (iii) there is often no information about what the outcome would have been if another decision had been taken; and (iv) most important decisions are unique and therefore provide little opportunity for learning (see Einhorn and Hogarth 1978). The conditions for organizational learning are hardly better. Learning surely occurs, for both individuals and organizations, but any claim that a particular error will be eliminated by experience must be supported by demonstrating that the conditions for effective learning are satisfied.

Finally, it is sometimes argued that failures of rationality in individual decision making are inconsequential because of the corrective effects of the market (Knez, Smith and Williams 1985). Economic agents are often protected from their own irrational predilections by the forces of competition and by the action of arbitrageurs, but there are situations in which this mechanism fails. Hausch, Ziemba, and Rubenstein (1981) have documented an instructive example: the market for win bets at the racetrack is efficient, but the market for bets on place and show is not. Bettors commonly underestimate the probability that the favorite will end up in second or third place, and this effect is sufficiently large to sustain a contrarian betting strategy with a positive expected value. This inefficiency is found in spite of the high incentives, of the unquestioned level of dedication and expertise among participants in the racetrack markets, and of obvious opportunities for learning and for arbitrage.

Situations in which errors that are common to many individuals are unlikely to be corrected by the market have been analyzed by Haltiwanger and Waldman (1985) and by Russell and Thaler (1985). Furthermore, Akerlof and Yellen (1985) have presented their near-rationality theory, in which some prevalent errors in responding to economic changes (e.g., inertia or money illusion) will (i) have little effect on the individual (thereby eliminating the possibility of learning), (ii) provide no opportunity for arbitrage, and yet (iii) have large economic effects. The claim that the market can be trusted to correct the effect of individual irrationalities cannot be made without supporting evidence, and the burden of specifying a plausible corrective mechanism should rest on those who make this claim.

The main theme of this article has been that the normative and the descriptive analyses of choice should be viewed as separate enterprises. This conclusion suggests a research agenda. To retain the rational model in its customary descriptive role, the relevant bolstering assumptions must be validated. Where these assumptions fail, it is instructive to trace the implications of the descriptive analysis (e.g., the effects of loss aversion, pseudocertainty, or the money illusion) for public policy, strategic decision making, and macroeconomic phenomena (see Arrow 1982; Akerlof and Yellen 1985).

References

Akerlof, G. A., and Yellen, J. 1985. Can small deviations from rationality make significant differences to economic equilibria? *American Economic Review* 75:708-20.

Allais, M. 1953. Le compartement de l'homme rationnel devant le risque: Critique des postulates et axiomes de l'Ecole Americaine. *Econometrica* 21:503-46.

Allais, M. 1979. The foundations of a positive theory of choice involving risk and a criticism of the postulates and axioms of the American School. In M. Allais and O. Hagen (eds.), *Expected Utility Hypotheses and the Allais Paradox*. Dordrecht:Reidel.

Arrow, K. J. 1982. Risk perception in psychology and economics. *Economic Inquiry* 20:1-9.

Bazerman, M. H. 1983 Negotiator judgment. *American Behavioral Scientist* 27:211-28.

Bell, D. E. 1982. Regret in decision making under uncertainty. *Operations Research* 30:961-81.

Bishop, R. C., and Heberlein, T. A. 1979. Measuring values of extra-market goods: Are indirect measures biased? *American Journal of Agricultural Economics* 61:926-30.

Chew, S. H. 1983. A generalization of the quasilinear mean with applications to the measurement of income inequality and decision theory resolving the Allais paradox. *Econometrica* 51:1065-92.

Chew, S. H., and MacCrimmon, K. 1979. Alpha utility theory, lottery composition, and the Allais paradox. Working Paper no. 686. Vancouver: University of British Columbia.

Clark, H. H., and Clark. E. V. 1977. *Psychology and Language.* New York: Harcourt Brace Jovanovich.

Einhorn, H, J. and Hogarth, R. M. 1978. Confidence in judgment: Persistence of the illusion of validity. *Psychological Review* 85:395-416.

Ellsberg, D. 1961. Risk ambiguity, and the Savage Axioms. *Quarterly Journal of Economics* 75:643-69.

Eraker, S. E., and Sox, H. C. 1981. Assessment of patients' preferences for therapeutic outcomes. *Medical Decision Making* 1: 29-39.

Fischhoff, B. 1983. Predicting frames. *Journal of Experimental Psychology: Learning Memory and Cognition* 9:103-16.

Fishburn, P. C. 1982. Nontransitive measurable utility. *Journal of Mathematical Psychology* 26:31-67.

Fishburn, P. C. 1983. Transitive measurable utility. *Journal of Economic Theory* 31:293-317.

Fishburn, P. C. 1984. SSB utility theory and decision making under uncertainty. *Mathematical Social Sciences* 8:253-85.

Fishburn, P. C. 1985. Uncertainty aversion and separated effects in decision making under uncertainty. Working paper. Murray Hill, N.J.: AT & T Bell Labs.

Fishburn, P. C., and Kochenberger, G. A. 1979. Two-piece von Neumann-Morgenstern utility functions. *Decision Sciences* 10:503-18.

Grether, D. M. 1980. Bayes rule as a descriptive model: The representativeness heuristic.*Quarterly Journal of Economics* 95:537-57.

Grether, D. M., and Plott, C. R. 1979. Economic theory of choice and the preference reversal phenomenon. *American Economic Review* 69:623-38.

Hagen, O. 1979. Towards a positive theory of preferences under risk. In M. Allais and O. Hagen (eds.), *Expected Utility Hypotheses and the Allais Paradox*. Dordrecht: Reidel.

Haltiwanger, J., and Waldman, M. 1985. Rational expectations and the limits of rationality: An analysis of heterogeneity. *American Economic Review* 75:326-40.

Hammond, P. 1985. Consequential behavior in decision trees and expected utility. Institute for Mathematical Studies in Social Sciences Working Paper no. 112. Stanford, California: Stanford University.

Hansson, B. 1975. The appropriateness of expected utility model. *Erkenntnis* 9:175-93.

Hausch, D. B.; Ziemba, W. T.; and Rubenstein, M. E. 1981. Efficiency of the market for racetrack betting. *Management Science* 27:1435-52.

Hershey, J. C., and Schoemaker, P. J. H. 1980. Risk taking and problem context in the domain of losses: An expected utility analysis. *Journal of Risk and Insurance* 47:111-32.

Kahneman, D.; Knetsch, J.L.; and Thaler, R. H. 1986. Fairness and the assumptions of economics. *Journal of Business* 59(4).

Kahneman, D.; Knetsch, J. L.; and Thaler, R. In press. Perceptions of fairness: Entitlements in the market. *American Economic Review*

Kahneman, D., and Tversky, A. 1979. Prospect theory: An analysis of decision under risk. *Econometrica* 47:263-91.

Kahneman, D., and Tversky, A. 1982. The psychology of preferences. *Scientific American* 246:160-73.

Kahneman, D., and Tversky, A. 1984. Choices, values, and frames. *American Psychologist* 39:341-50.

Karmarkar, U.S. 1978. Subjectively weighted utility: A descriptive extension of the expected utility model. *Organizational Behavior and Human Performance* 21: 61-72.

Knetsch, J. L., and Sinden, J. A. 1984. Willingness to pay and compensation demanded: Experimental evidence of an unexpected disparity in measures of value. *Quarterly Journal of Economics* 99:507-21.

Knez, P.; Smith, V. L.; and Williams, A. W. 1985. Individual rationality, market rationality and value estimation. *American Economic Review: Papers and Proceedings* 75:397-402.

Loomes, G., and Sugden, R. 1982. Regret theory: An alternative theory of rational choice under uncertainty. *Economic Journal* 92:805-24.

Luce, R. D., and Krantz, D. H. 1971. Conditional expected utility. *Econometrica* 39:253-71.

Luce, R. D., and Narens, L. 1985. Classification of concatenation measurement structures according to scale type. *Journal of Mathematical Psychology* 29:1-72.

Machina, M. J. 1982 "Expected utility" analysis without the independence axiom. *Econometrica* 50:277-323.

McNeil, B. J.; Pauker, S. G.; Sox, H. C. Jr.; and Tversky, A. 1982. On the elicitation of preferences for alternative therapies. *New England Journal of Medicine* 306:1259-62.

March, J. G. 1978. Bounded rationality, ambiguity, and the engineering of choice. *Bell Journal of Economics* 9:587-608.

Markowitz, H. 1952. The utility of wealth. *Journal of Political Economy* 60:151-58.

Nelson, R. R., and Winter, S. G. 1982. *An evolutionary Theory of Economic Change*. Cambridge, Mass.: Harvard University Press.

Payne, J. W.; Laughhunn, D. J.; and Crum, R. 1980. Translation of gambles and aspiration level effects in risky choice behavior. *Management Science* 26:1039-60.

Quiggin, J. 1982. A theory of anticipated utility. *Journal of Economic Behavior and Organization* 3:323-43.

Raiffa, H. 1968. *Decision Analysis: Introductory Lectures on Choices under Uncertainty* Reading, Mass: Addison-Wesley.

Russell, T. and Thaler, R. 1985. The relevance of quasi-rationality in competitive markets. *American Economic Review* 75:1071-82.

Savage, L. J. 1954. *The Foundations of Statistics.* New York:Wiley.

Schelling, T. C. 1981. Economic reasoning and the ethics of policy. *Public Interest* 63:37-61.

Schmiedler, D. 1984. Subjective probability and expected utility without additivity. Preprint Series no. 84. Minneapolis: University of Minnesota, Institute for Mathematics and Its Applications.

Schumpeter, J. A. 1954. *History of Economic Analysis.* New York: Oxford University Press.

Segal, U. 1984. Nonlinear decision weights with the independence axiom. Working Paper in Economics no. 353. Los Angeles: University of California, Los Angeles.

Simon, H. A. 1955. A behavioral model of rational choice. *Quarterly Journal of Economics* 69:99-118.

Simon, H. A. 1978. Rationality as process and as product of thought. *American Economic Review: Papers and Proceedings* 68:1-16.

Slovic, P.; Fischhoff, B.; and Lichenstein, S. 1982. Response mode, framing, and information processing effects in risk assessment. In R. M. Hogarth (ed.), *New Directions for Methodology of Social and Behavioral Science: Question Framing and Response Consistency.*San Francisco: Jossey-Bass.

Slovic, P., and Lichtenstein, S. 1983. Preference reversals: A broader perspective. *American Economic Review* 73:596-605.

Slovic, P., and Tversky, A. 1974. Who accepts Savage's axiom? *Behavioral Science* 19:368-73.

Smith, V. L. 1985. Experimental Economics: Reply. *American Economic Review* 75:265-72.

Stigler, G. J., and Kindahl, J. K. 1970. *The Behavior of Industrial Prices.* New York: National Bureau of Economic Research.

Thaler, R. H. 1980. Towards a positive theory of consumer choice. *Journal of Economic Behavior and Organization* 1:39-60.

Tversky, A. 1969. Intransitivity of preferences. *Psychological Review* 76:105-10.

Tversky, A. 1977. On the elicitation of preferences: Descriptive and prescriptive considerations. In D. E. Bell, R. L. Keeney, and H. Raiffa (eds.), *Conflicting Objectives in Decisions.* New York: Wiley.

Tversky, A., and Kahneman, D. 1981. The framing of decisions and the psychology of choice. *Science* 211:453-58.

Tversky, A., and Kahneman, D. 1983. Extensional versus intuitive reasoning: The conjunction fallacy in probability judgment. *Psychological Review* 90:293-315.

von Neumann, J., and Morgenstern, O. 1944. *Theory of Games and Economic Behavior.* Princeton, N.J.: Princeton University Press.

Weber, R. J. 1982. The Allais paradox, Dutch auctions, and alpha-utility theory. Working paper. Evanston, Ill.: Northwestern University.

Yaari, M. E. 1984. Risk aversion without decreasing marginal utility. Report Series in Theoretical Economics. London: London School of Economics.

HABITUAL DOMAIN ANALYSIS FOR EFFECTIVE DECISION MAKING[1]

Y. Shi
Graduate Assistant

P.L. Yu
C. A. Scupin Distinguished Professor
School of Business
University of Kansas
Lawrence, Kansas 66045, USA

Contents

---

1. The research has been partially supported by NSF Grant No. IST - 8418863

NATO ASI Series, Vol. F56
Multiple Criteria Decision Making and Risk Analysis
Using Microcomputers. Edited by B. Karpak and S. Zionts

1. Introduction

Human beings as complex living systems are constantly making decisions. Many decision making problems are routine and can be analyzed by mathematical models; and many more are nontrivial which challenge the traditional formal analysis. To successfully solve the nontrivial decision problems, we need to more fully understand human behavior mechanism and encompass the mechanism in our analysis. Let us consider the following example adapted from De Bono [5].

Example 1.1 A retiring chairman of a corporation invited to his ranch two finalists (A and B) from whom he would select his heir. A and B were given a black and white horse respectively. The chairman gave a course for the horse race and said, "Starting at the same moment, whoever's horse is slower in completing the course will be the next chairman."

Both candidates were puzzled because horse racing was habitually known to compete speed and the faster should win. The chairman's game was totally out of their comprehension.

Finally, A, getting out his habitual ways of thinking (habitual domains), jumped on B's horse and rode as fast as he could to the finish line. When B realized what was going on, it was too late. Naturally, A became the new chairman.

Observe that this nontrivial problem cannot be solved if A and B both stick to their fixed habitual domains (HDs). It was A who could first jump out of the HD and win the competition.

The above example illustrates that nontrivial decision problems may not be easily mathematically formulated. It is the concept of habitually domains which can successfully be used to analyze and solve the problems. It is due to the chairman's understanding about human behavior that he could design such a good competition. Indeed, if the chairman did not have a great HD, how could he manage his subordinates and survive in tough business competition?

In this paper, we shall outline (i) the general features of human behavior mechanism and its general tendencies (Section 2), which leads to (ii) habitual domain analysis (Section 3) and (iii) effective decision making (Section 4). Some interesting research problems are listed in Section 5.

2. Behavior Mechanism

2.1. A Dynamic Model of Behavior Mechanism

Using Flowchart 1, we sketch a dynamic model of behavior mechanism as follows. For more details the reader is referred to [16] or Chapter 9 of [17].

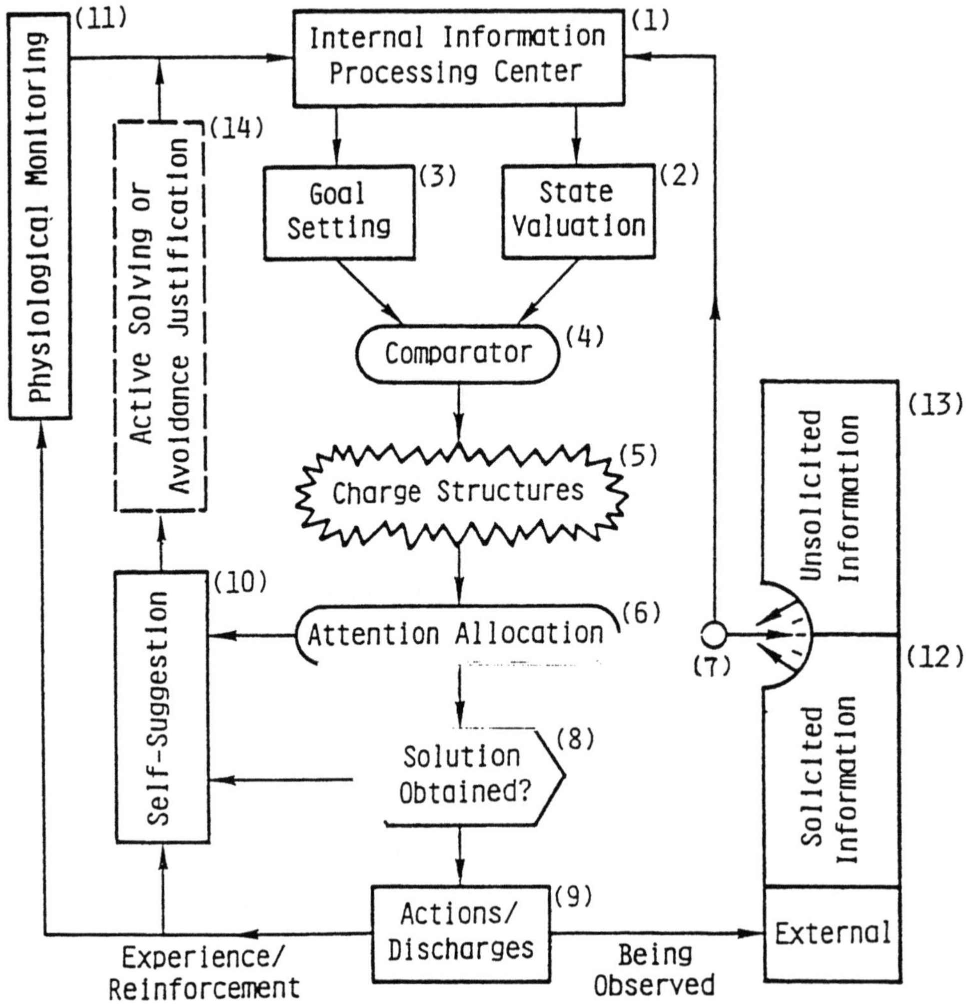

Flowchart 1 Behavior Mechanism

(I) Each individual is endowed with an internal information processing and problem solving capacity. Like a "super computer," the brain encodes, stores and retrieves ideas and messages, using the pattern circuits of excited neural cells. The stored messages can be dynamically restructured so as to most efficiently retrieve needed relevant information. Using association and analogy, the brain interprets arriving inputs based on the stored memory or learned knowledge. (See Box 1); also Hypotheses 1-4 of [16, 17] for details.) The arriving events can come from self-suggestion (Box 10), physiological monitoring (Box 11) and/or external inputs (Boxes 12, 13 and 7).

(II) Each human being has a set of goals or equilibrium points to reach and maintain (<u>goal setting</u>). A list of goals is given in Table 1. The status or state of each goal is constantly (consciously or unconsciously) monitored (<u>state valuation</u>). When a significant "unfavorable" deviation between the <u>perceived</u> value and the <u>ideal</u> value exists, each goal can produce a "charge" (a tension or urge). The totality of the charges produced by all goals which form a hierarchical structure is called <u>charge structure</u>. This charge structure can change rapidly with time, information inputs, and one's psychological state and condition. (See Boxes 2-5; also Hypotheses 5-6 of [16, 17].) Note that when a goal (e.g., health) is maintained at the tolerable interval of the ideal, without other effort, the goal may produce very little or no charge and will be neglected.

Table 1. A Structure of Goal Functions

(i)	Survival and Security: physiological health (proper blood pressure, body temperature and balance of biochemical states); proper level and quality of air, water, food, heat, clothes, shelter and mobility; safety; acquisition of money and other economic goods;
(ii)	Perpetuation of the Species: sexual activities; giving birth to the next generation; family love, health and welfare;
(iii)	Self-Importance Feeling: self-respect and self-esteem; esteem and respect from others; power and dominance; recognition and prestige; achievement; creativity; superiority; accumulation of money and wealth; giving and accepting sympathy and protectiveness;
(iv)	Social Approval: esteem and respect from others; friendship; affiliation with (desired) groups; conformity with group ideologies, beliefs, attitudes and behaviors; giving and accepting sympathy and protectiveness;
(v)	Sensuous Gratification: sexual; visual; auditory; smell; taste; tactile;
(vi)	Cognitive Consistency and Curiosity: consistency in thinking and opinions; exploring and acquiring knowledge, truth, beauty and religion;
(vii)	Self-Actualization: ability to accept and depend on the self, to cease from identifying with others, to rely on one's own standards, to aspire to the "ego-ideal" and to detach oneself from social demands and customs when desirable.

(III) Attention Allocation (Box 6) is defined as a human's time allocation of his internal information processing and problem solving capacity over various events and activities. The time could be measured in milliseconds. The purpose of attention is to release the charges in the most efficient way. These ways involve: (a) actions and discharges when solutions are obtained (Boxes 8-9); (b) acquiring external information (Boxes 7, 12, and 13); or (c) self-suggestion for internal thinking, justification and rationalization (Box 10). All of these functions feed back to the internal information processing center. (See Hypothesis 6 of [16,17].)

(IV) Each event or problem can be associated with a set of goals. Its significance can be defined in terms of the remaining charge structures when the event is removed. According to lexicographical ordering, the significance of events can be compared. The most significant event or problem will command one's attention. As charge structures rapidly change, the attention can switch rapidly from one event to other events. (See Hypothesis 6 of [16,17].)

(V) When there is a set of alternatives for discharge, the one which can most reduce the current charge structures will be chosen for discharge. This is called the least resistance principle. The remaining charge is a measure of resistance to the total discharge. (See Hypothesis 7 of [16,17].) Note that (IV) and (V) are similar. Both of them are complex dynamic optimization problems for our brain and mind.

(VI) When the existing set of alternatives cannot adequately reduce the level of charge structures, people have two possible behavior tendencies (Box 14): (A) Active problem solving (without changing the ideal value of the goals, the charge is not released until the ideal is reached), and (B) Avoidance justification (using self-suggestion to lower the ideal value or significance of the problem as to reduce the charge level). Note that active problem solving transforms high level of charge into force, drive or <u>motivation</u>; but avoidance justification simply diffuses or dissipates the charges. In the latter case, charges are not equal to force, drive or motivation.

(VII) All functions/components of the flow chart are interconnected. Through time, they interact with one another. For instance, once an action is taken (Box 9), say publishing an article, the event and its consequence will be observed and registered in the decision maker's memory (Box 1) and also will likely be observed and interpreted by other parties (Boxes 12 and 13) which may, in turn, react upon the original decision maker (Boxes 7 and 11). (See Hypothesis 8 of [16,17] for information inputs.)

<u>Remark 2.1</u> Observe that according to the model, there are a number of factors or functions, as showed in the boxes of Flowchart 1, which from the basis of decision making. Unfortunately, these factors or functions are usually very difficult to be described by mathematical models. However, with suitable effort and attention we may still be able to know how they are working in our own behavior.

The dynamic mechanism, although changing with time and with our psychological states, can become stabilized and unless extraordinary events occurs, each human being will have stable habitual patterns for processing information. We shall capture this observation by <u>habitual domains</u> which will be discussed in Section 3.

2.2. <u>Behavioral Tendencies</u>

The voluminous literature of social psychology has contributed to the understanding of human social behavior ([6,19] are good examples).

In this section, we shall list some interesting macro observations which can complement and strengthen our understanding of the human behavior mechanism. All listed observations appear in [6]. They are also discussed in great detail in the most textbooks with documentation. The observations are closely related to the micro mechanism (the dynamic model of behavior mechanism discussed in previous subsection). For a discussion of such relationship, we refer the reader to [6].

All the observations listed below should be carefully interpreted by tacitly adding "in general" or "there is a tendency."

Social Comparison Theory

"People have a drive to evaluate themselves and in the absence of objective non-social means, they evaluate themselves by comparison with other people."

This theory is closely related to goal setting and state valuation of the previous subsection. Many goal functions (refer to Table 1) are difficult to evaluate by ourselves. Through comparison with other people we know "better" where we stand with respect to the goals.

As a consequence of this theory and the goal of self-importance feeling, people may resent those capable of better performance than themselves and may look down on those with worse performance.

Halo Effect

"Most judgments of other people are made primarily in terms of good and bad. Then all their other qualities are deduced from this decision. One who is labeled good is surrounded with a positive aura and all good qualities are attributed to him. Conversely, one who is labeled bad is seen as having all bad qualities."

This observation is clearly related to the analogy/association rule of interpreting arriving events. As a consequence, we see that one who has the "right" posture can command favorable responses and the one who has the "wrong" posture may suffer from unfavorable reactions.

Projection Effect (Assumed Similarity)

"There is a strong tendency for people to assume that others are similar to them."

Thus, if one likes large parties, he tends to assume that other people like large parties; if he is aggressive, he assumes other people are also aggressive; and so on. This tendency is related to the analogy/association rule for interpreting arriving events and social comparison behavior.

Note that there are two implications from this observation. First, an individual may rate another more similar to himself than he actually is. That is, he distorts the other's personality to make it more like his own (self-suggestion). Second, this distortion may be so great that his rating of the other person corresponds more to his own personality (as he sees it) than to the other's personality (as other sees it).

Proximity Theory

"People who live near each other are much more likely to be friends and like each other than those who live far apart."

When people live near each other, they have more opportunity to meet and understand each other. Each one will have a better image of the other than in the absence of interaction. Predictability, familiarity and reinforcement will then help the development of friendship.

Reciprocation Behaviors

"One tends to like (or dislike) people whom he knows like (or dislike) him."

This behavior is related to rewarding. People like others who reward them or who are associated with pleasant experiences. Conversely, people dislike others who punish them or who are associated with unpleasant experience. Clearly this behavior is related to memory structure, association rule, goal setting, etc. Note that reciprocation also work for the feeling of cooperation, respect, helpfulness, hatred, etc.

Similarity Effect

"There is a strong tendency for people to like others who are similar to them. Moreover, society generally assume this to be true."

This may be due to the need for cognitive consistency. People want things to fit together and to be logical and harmonious, and this holds for their own beliefs, cognitions, thoughts and feelings. This need together with analogy/association rule for interpreting events can produce the similarity effect.

Scapegoating Behaviors (Displacement of Aggression)

"When the source of frustration or annoyance cannot be attacked, people have a tendency to express aggression against a substitute for the original source."

This behavior can be expressed in terms of charge structures and their releases by the least resistance principle.

Responsibility Diffusion or Deindividualization in Group Behaviors

"People to some extent lose their personal sense of responsibility when they are in a group. Instead of feeling, as they usually do, that they, personally, are morally accountable for their actions, groups members somehow share the responsibility with one another, and none of them feel it as strongly as he would if he were alone."

When people are in a group, they tend to feel as a part of the group. Responsibility and consequences of their individual actions are perceived to be shared among the group members and less identifiable. As a consequence, people in a group are more likely to take high-risk activities or to engage in acts they would not perform alone. The degree of deindividualization will certainly depend on the individual's charge structures and his commitment and relationship to the group.

3. Habitual Domain Analysis

3.1. Introduction

It has been recognized that each human being has habitual ways to respond to stimuli. Those observations of behavior listed in the previous section and conditioned or programmed behaviors are some of the descriptions of these habitual responses. Observe that each one of us has habitual ways of eating, dressing and using special phrases of languages. Some habitually emphasize economical gains, while others habitually emphasize social reputation. Some are habitually persistent in their goal pursuit, while others habitually change their objectives. Some are habitually optimistic and see the bright side, while others are habitually pessimistic and see the dark side. Some habitually pay attention to the details; while others only to the surface.

Similarly, in the perpetuation of species, life, safety, curiosity satisfaction and self-actualization, etc., each person has his own habitual way of thinking, acting and requirements.

In order to illustrate the habitual domains further, let us use an example.

Example 3.1 (Adapted from [1]) Alinsky was a social activist in 1960. In order to teach his followers a basic lesson of social movement, one day he and four followers, not in coat or tie, walked out of the Bilmore Hotel in Los Angeles. Alinsky held a ten dollar bill and tried to give it away to strangers by saying, "Here, take this." The following are some different responses:

A. "I'm sorry, I don't have change."

B. "I'm sorry, I don't have any money on me right now."

C. "I am not that kind of a girl, and if you don't get away from here, I'll call a cop!"

D. "I do not come that cheap."

E. "What kind of con game is this?"

Note that with respect to the same stimuli of the ten dollar bill, different people react differently. Indeed each response is a reflection of a habitual way of thinking and judging the situation.

Those habitual ways of perceiving, thinking, responding and acting can be captured by the concept of habitual domains (HD). Flowchart 1 illustrates that HD involve self-suggestions, external information inputs, physiological monitoring, goal setting, state valuation, charge structures, attention and discharges. They also involve encoding, storing, retrieving and interpretation mechanisms. When a particular aspect or function is emphasized, it will be designated as "HD on that function." Thus, HD on self-suggestion, HD on charge structures, HD on attention, etc. all follow. When the responses to a particular event are of interest, we can designate them as "HD on the responses to that event." Thus, HD on job seeking, house purchasing, dealing with friends, etc. also follow. In the following discussion, HD are used without specifying the corresponding functions or events, as only the general properties of HD will be discussed.

3.2. Elements of Habitual Domains

A habitual domain at time t, denoted by HD_t, is defined as the collection of the following:

(i) PD_t (potential domain): The collection of ideas/actions that can be potentially activated at time t;

(ii) AD_t (actual domain): The set of ideas/actions that is actually activated at time t;

(iii) AP_t (activation probability): The activation probability or confidence structure at time t which indicates the possibility for a subset of ideas/actions in PD_t to be in AD_t;

(iv) $R(I_t,O_t)$ (reachable domain): The reachable (attainable) set of ideas/actions from the initial set of ideas I_t through the set of operators O_t.

$R(I_t,O_t)$ shall be described more extensively shortly. Observe that $AD_t \subset PD_t$. The relationship between AD_t and PD_t is similar to that between the realized value and the sampling space of a random variable.

The set AD_t will be affected by the charge structures and attention at time t. The probability or confidence level (AP_t) for a set of ideas/actions to be activated will depend on how strongly the ideas/actions have been encoded and how easily the ideas/actions can be retrieved from memory storage. It is also closely related to the set of initially activated ideas around the time t.

Now consider reachable domains. Through self-suggestion and/or external information inputs, one idea or a set of ideas can be used to stimulate or generate other ideas. This suggests that a set of operators exists, defined on the subsets of PD_t, which generate ideas in PD_t from the subsets of PD_t. As an example, suppose that one is interested in investing a fixed amount of cash in stocks A and B. The concept (the operator) that any portfolio (a convex combination) of A and B would also be of interest will expand the alternative set of A and B into the set of all convex combinations of A and B. Note, the operators are also elements of PD_t.

Let I_t be a set of ideas with $I_t \subset PD_t$, and let O_t be a set of operators which generate ideas in PD_t from subsets of PD_t. Then $R(I_t,O_t)$, the reachable domain, is the set of ideas/actions that can be reached (or attained) from I_t and O_t. More precisely, $R(I_t,O_t)$ is the set of ideas/actions that can be cumulatively generated by any sequence of operators from the set of O_t which act on I_t and the resulting ideas/actions from the operations. As an example, let $I_t = \{0,1\}$ and $O_t = \{+\}$ ("+" is the ordinary mathematical addition symbol). Then $R(I_t,O_t)$ is the set of all nonnegative integers. If $O_t = \{+,-\}$, then $R(I_t,O_t)$ is the set of all integers. For athletes, the idea of becoming an Olympic star can generate "glory, pride, wealth, hard work, discipline . . ." as a subset of the reachable domain.

Potential and reachable domains are closely related. We say that $\{I_t, O_t\}$ is a <u>generator</u> of PD_t iff $PD_t = R(I_t, O_t)$, and $\{I_t, O_t\}$ is a <u>basis</u> for PD_t iff $\{I_t, O_t\}$ is a generator of PD_t <u>and</u> no strict subset of $\{I_t, O_t\}$ is a generator of PD_t.

Finally we observe that HD_t, as defined, is dynamic and changes with time. All four elements of HD_t can evolve with time: this makes analysis of HD_t fairly difficult. Fortunately, it can be shown that the HD_t can reach a stable state. This will be sketched in the next subsection.

3.3. <u>Existence of Stable Habitual Domains</u>

In [4], we show the PD_t and P_t depends on the activation of neural pattern circuits. With suitable conditions, the pattern circuits will satisfy a system of differential equations, which have stable steady solutions. The interested reader is referred to [4] or [17]. Here we shall derive the stability of PD_t, and discuss some important implications of stable HD_t.

The existence of stable state of PD_t can be roughly derived as follows. As each human being learns and experiences, his PD_t grows with time, but at a decreasing rate because the probability for an arriving idea to be new with respect to PD_t is getting smaller as PD_t gets larger. Thus, unless unexpected extraordinary events arrive PD_t will reach its stable state. Once such a state is reached, habitual ways of thinking and responses to stimuli can be expected. If extraordinary events do not arrive too often, as is usual, then PD_t will be in its stable state. Thus habitual ways of thinking and acting will be manifested most of the time. This observation was the main motivation to use "habitual" as the adjective.

In order to derive a more precise theorem, let a_t be the number of additional new ideas or concepts acquired during the time interval $(t-1,t]$. Then $PD_t = \sum_{s=0}^{t} a_s$.

Note that $a_t \geqq 0$ and $\{a_t\}$ may not be monotonic. It may have some patterns of periodicity, as a result of bio-rhythm or rhythms of the environment (such as day and night).

<u>Theorem 1</u> (Imbedding Theory). Assume that there exists a subsequence $\{a_{s_k} | k=1,2,\ldots\}$ of $\{a_t\}$ and constants D and M, such that

(i) $\sum_k a_{s_k} \to D$

(ii) $\sum_{i=1}^{s_{k-1}-s_k} a_{s_k} + i \leqq M\, a_{s_k}$

Then $\sum_{t=0}^{\infty} a_t$ converges.

<u>Proof</u>:

$$\sum_{t=0}^{\infty} a_t = \sum_{t=0}^{s_1} a_t + \sum_{i=1}^{s_2-s_1} a_{s_1+i} + \sum_{i=1}^{s_3-s_2} a_{s_2+i} + \cdots$$

$$\leqq \sum_{t=0}^{s_1} a_t + M(a_{s_1} + a_{s_2} + \cdots)$$

$$\leqq \sum_{t=0}^{s_1} a_t + MD < \infty$$

Once PD_t and P_t reach their stable states, we can expect AD_t to occur at some regularity. In such stable states of HD_t we can expect habitual ways of thinking, responding, and reacting to stimuli and events to occur. Thus, personality, attitude, conditioned or programmed behavior, will be formed for each individual. Such formation has a great impact on decision making styles and optimization theories.

4. Effective Decision Making

4.1. Cores of Habitual Domain and A Classification of Decision Problems

As discussed in previous sections, understanding the HDs of ourself and other people involved are important for effective decision making. To make our study more concrete, we shall sketch a new concept -- the cores of HDs (see [18] for details).

Roughly speaking, the core of the HD is the set of ideas or concepts that are very strongly implanted in the HD and can be easily retrieved to interpret the arriving events or to perform the needed work. As a core is a concept on how strongly the ideas are encoded and how easily retrieved, while a generator or basis is a concept on those ideas that can potentially activated, we notice that the core can usually just be a proper subset of a generator or a basis. For example, when we talk about relatives and friends, the set of {the parents, the wife or husband, the sons and daughters, the business colleagues, the neighbors} can be a core; but the set cannot be a generator or a basis for all relatives and friends. New ideas can be implanted into HD and become elements of the core.

In order to make the concept of core precise, let I be a given idea and let E be a particular event or problem under consideration.

Define the activation density of idea I in event E at time t by:

$$a_t(I,E) = \text{Prob}_t[I \text{ activates} | E \text{ gets attention}] \qquad (4.1)$$

which is the conditional probability for I to be activated given that our attention is paid to E at time t.

Note that the activation density is a function of time and $0 \leq a_t(I,E) \leq 1$. Now we can define the α-core of HD with respect to E at time t, denoted by $C_t(E,\alpha)$ to be the set of ideas or concepts that have activation density exceeding or equal to α. That is,

$$C_t(E,\alpha) = \{I | a_t(I,E) \geq \alpha\}. \qquad (4.2)$$

Note that if $I \in C_t(E,\alpha)$, with high value of α, (say $\alpha \geq 0.9$), then I is an active and very frequently retrieved idea when E catches our attention. Also note the $C_t(E,\alpha) \subset C_t(E,\beta)$ if $\alpha > \beta$.

In our later discussion, unless otherwise specified, whenever we talk about the core of HD we shall mean α-core with high value of α.

For convenience, given HD for a particular problem E, we shall use the following terminology.

By implanting an idea I into HD we mean putting idea I, perhaps subtly, into the subject's HD. At the time of implanting, the idea I is usually new or fairly new to the subject. That is, the activation density of I in E, $a_t(I,E)$, is zero or very low. As time goes by, the newly implanted idea can be rejected or accepted by HD. By accepting the idea I in HD, we shall mean that the idea I becomes an active element of HD with $a_t(I,E) > 0$. As I is repetitively used, its activation density will become larger and can eventually become an element of the core of HD. Thus when the problem E is attended, idea I will be habitually or automatically activated. A set of ideas $I = \{I_1,\ldots,I_k\}$ is said to become the habituation of event or problem E if I becomes the core of HD whenever E is attended.

Furthermore, depending on the regularity and availability that ideas, concepts and operations are needed to successfully perform jobs or solve problems, we can roughly classify the decision problems into four classes:

Class I. Simple Routine Problems. In this class, the set of ideas, concepts and operations which are needed to successfully solve problems is almost fixed and well ordered. Assembly line work, typing, sleeping, eating, driving, etc. can belong to this class. With proper training, most people can acquire this set of ideas, concepts and operations and incorporate them into the core of HD for solving their problems. Thus, habituation of the order of operations becomes an important part in performing and solving the problems.

Class II. Mixed Routine Problems. To successfully execute or solve this class of problems, one needs to solve variable combinations of simple routine problems. The work of repair shops, receptionists, waitresses, clerks, and regular dinner preparation belong to this class. In this class, the set of ideas, concepts and operations needed to successfully solve the individual component problems are known, and can be acquired and habituated as the core of HD by proper training. However, the best order to perform the job may not be fixed. There are also time allocation problems over the individual component problems. One notices that the problems of this class are more complex than those of Class I. The core of HD in this class, in general, is an α-core with α smaller than that of Class I.

Class III. Fuzzy Problems. There are no sure set of ideas, concepts and operations that can guarantee the success of a solution, even though the set can be fuzzily known. Sales problems, advertisement problems, ordinary research and development problems, simple conflict problems with friends, problems of preparing for an important dinner party, etc. are some examples of this class of problems. There may exist a set of ideas, concepts and operations which are commonly but fuzzily known to be good for solving the problems. This set of ideas may be acquired and habituated through training and learning as to become a large proportion of the core of HD (i.e. α-core with high value of α) for solving the problems. However, this set alone is usually not adequate to surely and successfully solve the problems. There is always a need of new ideas or concepts, which may be acquired from the potential domain by active thinking and association, to successfully solve the problems. In general, α-core, with a large α, of the decision maker's HD is not adequate for solving the problems. In fact, rigid and inflexible HD (i.e. the α-core is almost fixed even if we lower the value of α) may prove to be detrimental to solving the fuzzy problems.

Class IV. Challenging Problems. These problems cannot be successfully solved by α-core of HD, no matter how low we reduce the value of α. These kinds of problems include innovative research and development problems which challenge the existing technical assumptions, market restructuring problems, complex conflict resolution problems, traumatic disasters, etc. The problems can be solved only by expanding and restructuring our HD. Fixed mindedness (fixed HD) usually becomes a major resisting force for solving challenging problems. In order to expand our HDs we need to go through the process of implanting, accepting and habituating.

Finally, we note that each one of us has solved some of the above four classes of problems, and someone's routine problems may be others' fuzzy or challenging problems. The above classification depends on each individual's HD, which results from his own experience and learning.

4.2. A General Principle of Effective Goal Setting

In this section we shall list a general principle of effective goal setting which is a key to effective decision making. The detailed applications of this principle, depending on the HDs of the performers and the kind of problems they are facing, can be found in [18].

Utilizing the concepts of Section 2 and 3, we can state the general principle of effective goal setting as follows:

"To select goal functions and their achievement levels, and to set up proper supportive systems as to create favorable working conditions, and relatively high levels of charges and confidence on the performers so that a continuous allocation of as much time as possible to perform the desired job-related operations in the most efficient and effective way is the least resistance method for the performers to release their charges (thus obtain their satisfactions)."

Note that the principle describes a complex optimal control system in which the control variables are goal setting (selecting goals and their achievement levels) and supportive systems; the state variables are working conditions, charge structures and confidence, which may be only partially observable; and the objectives are to maximize the attention allocation of time to job-related works, to maximize the efficiency/effectiveness of performing the works to release the performers' charge, and, perhaps, to maximize the favorability of the working environment.

Let us first integrate the above principle with the behavior mechanism as depicted in Flowchart 2. In the flow chart we assume that physiological conditions and external information inputs, except supportive systems, have been neutralized and do not create high levels of charges or disturbances on the performers. Thus they are put in the insignificant sub-box "others" of Box (4).

Note that in Flowchart 2, the boxes in the right column (i.e., Boxes (7) - (10)) are components of the supportive system; while those in the left column (i.e., Boxes (1) - (6)) are the modified "behavior mechanism." Let us explain the principle using Flowchart 2 as follows:

Boxes (1) - (2). Goal Setting and State Valuation. Although the functioning of these two boxes are as in Flowchart 1, they are subject to the influence and control of our purposeful goal setting and supportive system in addition to physiological monitoring and other external information inputs. In Box (1), "goal setting" has been used as a purposeful control on job performance as well as its general meaning of setting the ideal value of living goals beyond those related to job performance (i.e., including those listed in Table 1).

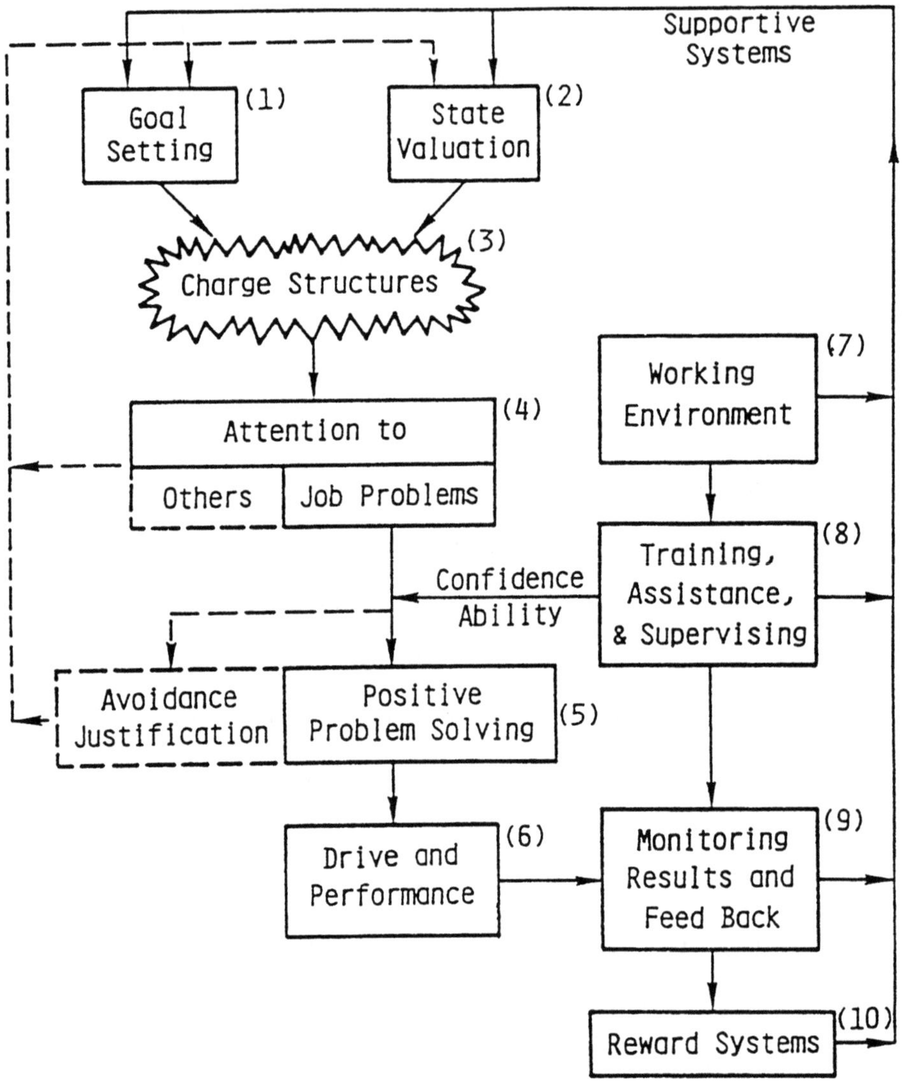
Supportive
Systems
Goal
Setting
(1)
State
Valuation
(2)
Charge Structures
(3)
Attention to
Others
Job Problems
(4)
Working
Environment
(7)
Training,
Assistance,
& Supervising
(8)
Confidence
Ability
Avoidance
Justification
Positive
Problem Solving
(5)
Drive and
Performance
(6)
Monitoring
Results and
Feed Back
(9)
Reward Systems
(10)

Flowchart 2.

Boxes (3) - (4). Charge Structures and Attention. In order to obtain attention allocation, the purposeful goal setting must be able to create the relatively highest level of charges on the performer's charge structures. That is, the charge of "not reaching the performance goals" is so much higher than that of "reaching the performance goals," that no life goal functions other than the performance goals can create a more significant charge. When the performance goals can create a sufficiently high level of charge, while all other life goals cannot do so, the performers can have a burning desire to achieve the set performance goals. To achieve this objective, we need to (A) create the rosy and promising picture of the state when the performance goals are accomplished (positive rewards); (B) let the performers realize the sad and depressing state when the performance goals are not accomplished (negative rewards or punishment); (C) set up barriers so that job-unrelated messages or disturbances would not reach the performers and create a high level of charge; and (D) develop good working environment in which the performers enjoy their work and in which accidents are unlikely to occur.

Note that in Box (4), there is a sub-box, Attention to Others, which includes physiological monitoring and external information inputs unrelated to job performance. The more attention paid to "others," the less attention paid to job-related work. It may represent a leakage of time and effort with respect to performance. Note that 100% of attention over a long period of time to job-related work is impossible, because men as living systems at least need to eat, sleep, etc. to release the high level of charges created by physiological goal functions (see [16,17] for further discussion).

Boxes (5) - (6). Positive Problem Solving or Avoidance Justification, and Drive and Performance. As mentioned in Section 2, positive problem solving is to work hard to accomplish the preset goal values, while avoidance justification is to lower the ideal goal value or the significance of the problems. Thus, after attention, the attitude of positive problem solving will produce "drive" and performance; while that of avoidance justification will usually not. Besides personality, confidence, hope and the ability to accomplish the ideal goal value in an acceptable time horizon plays an important role in deciding which attitude will be taken. Using supportive systems, especially training, assistance and supervising (Box 8), to increase the performer's confidence and ability is an important means to enhance the attitude of positive problem solving and to improve the performance.

Boxes (7) - (10). Supportive Systems These four boxes are some important devices for supportive systems. They help create or modulate charge structures as to make performers pay more attention to job-related problems, cultivate a positive problem solving attitude, improve confidence and ability to accomplish the ideal goal value, and to actually achieve the set goal value.

Note that by utilizing Flowchart 2, one could describe a dynamic system for the problem of effective goal setting. With proper assumptions such as additivity with respect to time, the dynamic system can even be expressed in terms of differential or integral equations (see [18]).

4.3. Typical Complex Decision Problems

Let us apply habitual domain analysis to the following typical complex problems or concepts.

(i) On high stake decision problems: Although the four decision elements: alternative, criteria, perceived outcomes of decisions and preference, can vary with time, information inputs, and psychological states of the decision maker, they can become stabilized (see [3] for an empirical study). Applications of optimization theories to high stake decision problems become feasible. Before the stabilization, a formal analysis using optimization theory is not fruitful. During the transition period we might be better off to let our HD_t open and expanded, allowing vigilant search for all relevant information on the four decision elements, making sure "good" alternatives are not overlooked. Management of the transition of HD_t can be a very challenging problem. It is especially so when our HD is not big enough as to comprehend the process of transitions.

(ii) On optimal solutions/retional decisions: As decision processes depend on HD_t so do the resulting optimal solutions. Since HD_t can vary over time (even though it can reach its stable states most of the time), optimal solutions will change with time. This occurs when the set of alternatives, the set of criteria, dimension of alternatives, perception of outcomes and preference change. This suggests that in dynamic settings, "time-optimality" is important and that an alternative perceived as optimal is valid only over a time horizon. Today's optimal solution does not imply that it will be optimal forever. As HD_t changes, it may become an inferior solution. See [14,15] for further details. Being aware of this fact we can avoid surprise over other people's (decision makers') "irrational" decision. After all, a decision is rational if and only if it is consistent with the decision maker's HD_t. Everyone's HD_t is unique. What we perceive as irrational may be very rational from other people's point of view (HD_t).

(iii) On gaming and conflict dissolution: Each player has a unique HD_t. Understanding our own HD_t and the opponents' HD_t is essential to winning competitive games or resolving conflicts. If we know our own HD_t but do not know the opponents' HD_t, we cannot confidently construct a winning strategy. Indeed we could lose the game entirely, like the Pearl Harbor disaster (see [9] for a detailed discussion of the mistake). If we do not know our own and the opponents' HD_t, very likely we would lose most games.

In partial cooperative and partial competitive games, like international trade or companies competing for market share and market volume with the same kinds of products, it might be desirable for the players to settle in peace and ensure that each one benefits. To maintain some stability the settlement must allow each player to declare a victory, or the terms of agreement must be a time-optimal solution from the point of view of each player's HD_t. Certainly this is not an easy task. Proposing new alternatives, creating new conceptions of the criteria and suggesting outcomes for the players to change their corresponding HD_t will become vital. Without a new set of compatible HD_t's, agreement can hardly be reached. Certainly, to successfully restructure HD_t we must first be aware of the existing HD_t of each player. Instead of using the framework of traditional game theory, such as described in [11,8], the second order game concepts of [15] using HD_t should be used to resolve nontrivial games or conflict.

As an illustration, let us consider:

Example 4.1. (Adapted from [13]) Once Liu-Pam (a Chinese Emperor about 2,000 years ago) and his troop was taken and surrounded by an overwhelming army commanded by the King and Queen of an earlier tribe of Mongolia. In this desperate crisis, Chen-Ping, a wise aide to the Emperor, derived an innovative strategy to dissolve the crisis.

"A famous artist was asked to paint an imaginatively charming and beautiful lady. The painting was secretly sent to Mongolia's queen with the message that the charming lady was going to be offered to the King for his victory and that she was so charming and attractive that the current queen would surely be abdicated."

On that night, the Queen, beloved and trusted by the King, subtly told the King that she had had a deadly dream the night before that both of them were killed in a battle with Chinese troops. She then successfully persuaded the King to retreat to cherish their love the next day!

Observe that to the Queen, the glory of conquering China was not as important as to have the King's love and that the thought of being abdicated was not fun at all! From the King's point of view, it was much better to be alive as a King than to be killed in battle. The glory of conquering China could of course never make up for the death of himself.

Note that the above brilliant strategy could not be obtained or studied by traditional game theory. Without a thorough understanding the HDs of all the players involved, the creative strategy could not be derived.

(iv) On career management: In a broader sense, each social organization (family, company, school, society, nation, etc.) can be regarded as a living entity and can have a habitual domain, HD_t. The habitual domain can also be stabilized. The person within the organization performing a variety of functions also has his own HD_t. The match of these habitual domains is important for considering career success and happiness. If the HDs are compatible and if they enhance each other, we may expect a fairly happy individual in the organization; otherwise, conflict and frustration can occur and adjustment must be made in order to keep the individual with the organization.

If we regard human beings as adapting entities to their environments, then choosing an organization for association becomes an optimal match problem between his HD_t and the organization's HD_t. Are there ideal organization HD_t to which an individual can adapt? Can he change the organizational HD_t or is it easier to change his? Should he leave an organization for a new one which reveals a more compatible HD_t? These are some important questions that each individual must address for career success and deserve a careful exploration. Some related literature can be found in Holland [7] and citations therein.

(v) On leadership: Each "group" or social organization as mentioned before has its own HD_t which depends, at least partially, on its members' HD_t. A good leader should have a clear picture of the HD_t of the group and the members. He should also have a clear picture of the ideal HD_t for the group. A leadership may be defined as moving the current HD_t of the group to an ideal HD_t for the group.

There are two extreme kinds of leadership. One is status-quo leadership which maintains the current HD_t. The other is revolutionary leadership which drastically changes the current HD_t to a new ideal HD_t.

Note that knowing the individual members' HD_t is very important for successful leadership. Indeed, a leader can best motivate members to move the group's HD_t when he can create a high "charge" on the individual members and show the way for them to successfully release the charge. Certainly, the perceived ideal group HD_t should be compatible with and able to enhance (as perceived by the members) the members' HD_t. Otherwise, resistance and conflicts will occur as to hinder the success of leadership. Some related literature can be found in [2].

4.4. Expansion of Habitual Domains

In Section 3.3, we have discussed how HD_t can become stabilized. Stabilized HD_t allows us to process information and to make decisions more effectively and efficiently on daily repetitive problems; it also makes our thinking and behavior predictable. In an extreme, our thinking and behavior may become routine and programmed, when our HD_t become sufficiently rigid and inflexible. As indicated in the previous subsections, in high stake decision problems, conflict resolution, career management and leadership, a rigid HD_t cannot successfully offer solution. Indeed, a rigid HD_t may become a major source of the problem. Developing a HD_t which can appreciate and understand other HD_t's is a very important step for us to enhance our ability to solve difficult problems and to have a happy career and life.

In this subsection we shall first describe some effective methods to expand and develop our HD_t. Then some technical concepts will be sketched to invite the interested reader to make contributions in this area.

The following methods can expand our HD_t:

(i) Expand HD_t by awareness and willingness to learn from other people. The largest resource for expanding our HD is other people's HDs. It is safe to say that everyone's HD is almost surely unique. That is, there is virtually zero probability that the HDs of any two persons can be identical. (See [16-17] for further discussion.) By being aware of the difference among HDs, we may begin to be willing and able to spend effort and time to encode and absorb those favorable ideas/operators so as to expand our HD.

Depending on the degree of cooperation, rivalry and intimacy, humans reveal their HDs to their partners, perhaps partially, consciously and/or unconsciously. Their revealed ideas/actions or operators may be accepted or absorbed because they are similar to those of the partners or because the partners make a special effort to do so. The ideas/actions or operators can also be rejected because they are strange to the partners and/or the partners either do not care or activate their self-suggestion to distort and avoid them. The acceptance or rejection will certainly depend on the partners' charge structure and attention, etc. Usually new ideas/actions or operators can be more easily learned and absorbed if they are similar to those which are already known than otherwise. Confronted with ideas which are outside our HD, we may have a tendency to reject them right away instead of taking time and effort to absorb them. This quick rejection to totally new ideas may prevent us from expanding our HD to a higher dimension.

In real life HDs of other people are not easily observed. Recall that the actual domain AD_t is a subset (perhaps very small) of PD_s. Most of the time only a portion of PD_t is observable, and without attention even the observable part of AD_t can be ignored and/or misinterpreted. The following are some important operators/ideas for awareness and absorption of other persons' HDs: (A) sincere appreciation and genuine interest in other persons' abilities, skills and well-being, (B) open minded frank discussion with other people, (C) an attitude that other persons' opinions/beliefs can be right or valid and (D) the willingness to think in terms of the other persons' interests.

(ii) Expand HD_t by considering one higher order of the systems within which we operate. For instance, in the stock market investment problem, one can regard the stock market as a subsystem of speculation markets which also include the bond market, option market, futures market, etc., which again is a subsystem of the money-making mechanism. By looking into a higher order system, we can usually broaden our mind and see other ideas/operators which we could not otherwise see.

(iii) Expand HD_t by periodically being aware of our existing HD and seriously questioning the preconceived assumptions and notions in the existing HD_t. Different assumptions and notions will likely produce different conclusions, just like different initial points and dynamics will produce different terminal points. This awareness and questioning can make us discover new ideas and expand our HD.

(iv) Expand HD_t by varying relevant parameters in the thinking process. For instance, in a house purchase decision, one can vary the parameter of how much he can borrow from the bank so as to expand the set of feasible houses to be purchased.

(v) Expand HD_t by actively using "association law." By actively trying to find the similarity and difference between two objects, one may be able to discover new ideas. For instance, in decision making, one may be able to expand the set of alternatives by deliberating the other three elements: criteria, perceived outcomes, and preference. Through this deliberation, some new alternatives may be derived.

(vi) Expand HD_t on specific problems by consulting with experts, previous similar events and books. This consultation can usually provide us new ideas which we could not obtain otherwise.

(vii) Retreat from the specific problem for a while when being trapped by our HD_t and when the above methods offer no help. Turning off the problem has an effect of turning off being trapped. Then start our generating process when our mind is refreshed again. New concepts and vital ideas can be obtained in an easier fashion when our mind is fresh and is not trapped by the previous thinking.

(viii) Change the environment or move into a new environment when all the above fail. Change of the environment provides new inputs into our system, which could stimulate our new thinking and expand our HD.

5. Conclusion

We have already sketched a dynamic mechanism of human behavior, habitual domain analysis, and effective decision making. More details of the related concepts can be found in Chapter 9 of [17], [16], and [18]. A number of research problems need to be further explored. For instance, how can we find effective ways to classify and identify HDs for a variety of decision situations? How can we know that the different HDs involved are compatible and enhancing each other, or are conflicting and destructive to each other? How can we change from an existing HD_t to an ideal HD_t^*? Certainly the applications of HD analysis to various problems as sketched in Section 4.3 still remain to be fine tuned and explored. All in all, to be successful in solving nontrivial decision problems, it is important for us to know our own HD, the HDs of other people involved in the decision making and/or conflicts, and the HDs of the environment.

References

1. Alinsky, S. D., Rules for Radicals, Vintage Books, 1971.

2. Burns, J. M., Leadership, Harper & Row, New York, 1978.

3. Chan, S. J., Park, C. W. and Yu, P. L., "High-Stake Decision Making -- An Empirical Study Based on House Purchase Processes," Human Systems Management, Vol. 3, pp. 91-106, 1982.

4. Chan, S. J. and Yu, P. L., "Stable Habitual Domains: Existence and Implications," Journal of Mathematical Analysis and Applications, Vol. 110, No. 2, pp.469-482, 1985.

5. De Bono, E., PO: A Device for Successful Thinking, Simon and Schuster, New York, New York, 1972.

6. Freedman, J. L., Carlsmith, J. M. and Sears, D. O., Social Psychology, Prentice-Hall, Englewood Cliffs, New Jersey, 1974.

7. Holland, J. L., Making Vocational Choices: A Theory of Careers, Prentice-Hall, Inc., New Jersey, 1973.

8. Isaacs, R., Differential Games, John Wiley and Sons, New York, 1965.

9. Janis, I. L. and Mann, L., Decision Making, A Psychological Analysis of Conflict, Choice and Commitment, The Free Press, New York, 1977.

10. Lindsay, P. H. and Norman, D. A., Human Information Processing, An Introduction to Psychology, Academic Press, New York, 1972.

11. Luce, R. D. and Raiffa, H., Games and Decisions, John Wiley and Sons, New York, 1967.

12. McCormack, M. H., What They Don't Teach You at Harvard Business School, Bantam Books, New York, New York 1986.

13. Su, M. C., Historical Records (Shih-Chi), Vol. 56, No. 26, Han-Dynasty (explained by Fei Yin, Song-Dynasty). (In Chinese.)

14. Yu, P. L., "Decision Dynamics with an Application to Persuasion and Negotiation," TIMS Studies in Management Sciences, Vol. 6, North-Holland Publishing Co., New York, 1977.

15. Yu, P. L., "Second-Order Game Problem: Decision Dynamics in Gaming Phenomena," Journal of Optimization Theory and Applications, Vol. 27, No. 1, January, 1979, pp. 147-166.

16. Yu, P. L., "Behavior Bases and Habitual Domains of Human Decision/Behavior -- An Integration of Psychology, Optimization Theory and Common Wisdom," International Journal of Systems, Measurement and Decisions, Vol. I, 1981, pp. 39-62.

17. Yu, P. L., Multiple Criteria Decision Making: Concepts, Techniques and Extensions, Plenum, New York, New York, 1985.

18. Yu., P. L., and Chien, I. S., "Habitual Domain Analysis and Effective Goal Setting for Better Performance," working paper, School of Business, University of Kansas, 1987, to appear in Human Systems Management.

19. Zimbardo, P. G. and Ruch, F. L., Psychology and Life, Scott, Foresman and Company, Glenview, Illinois, 1975.

GOAL SETTING and COMPROMISE SOLUTIONS

Y. Shi
Graduate Assistant

P.L. Yu
C. A. Scupin Distinguished Professor
School of Business
University of Kansas
Lawrence, Kansas 66045, USA

Contents

NATO ASI Series, Vol. F56
Multiple Criteria Decision Making and Risk Analysis
Using Microcomputers. Edited by B. Karpak and S. Zionts

1. Introduction

It has been recognized that each human being has a set of ideal goals to achieve and maintain. (See [11-12] for more detailed discussion.) Such goals include: (i) survival and security of the self: physiological health, safety, and freedom from danger; (ii) perpetuation of the species: sex, giving birth to the next generation, family love, and welfare; (iii) feelings of self-importance: self-respect and self-esteem, accumulation of wealth and power, dominance, recognition, prestige, achievement, etc.; (iv) social approval: esteem and respect from others, affiliation, conformity with a desired group, giving and accepting sympathy and protectiveness; (v) sensual gratification: sexual, visual, auditory, smell, taste, tactile; (vi) cognitive consistency and curiosity: consistency in thought and opinion, exploring and acquiring knowledge, truth, beauty, and religion; (vii) self-actualization: the ability to accept and depend on the self, to rely on one's own standard, to cease identifying with others.

Note that each goal in abstract is a function and the ideal (ideal value or the ideal goal) is a particular value of the (goal) function at which we want to achieve and maintain. The goal functions are under our continual monitoring, consciously or subconsciously.

Observe that some goals, such as accumulation of wealth, reputation, and achievement, are self-suggested, adjustable, and can be characterized by "more is better." Other goal settings, such as body temperature and other biochemical states of the body, are fairly fixed, without much room for self-adjustment or change. Up too much or down too much from the ideal setting is not desirable. The deviation can be disastrous. Once a perceived state (or value) has deviated from its targeted or ideal state (or value), a charge (tension or pressure) will be produced to prompt an action to reduce or eliminate the deviation. This behavior of taking actions, including adjustment of the ideal values, to move the perceived states to the targeted ideal states will be called goal-seeking behavior.

This goal-seeking behavior, which has been well documented in psychological literature (for instance, see [7]), has an important and pervasive impact on human decision making. In this paper, we shall focus on two concepts that define human goal-seeking behavior. The first one is related to satisficing models, which are treated in Section 2, the other is related to compromise models including goal programming, which are treated in Section 3. All of these are static models. For the dynamic version of goal-seeking behaviors we refer to Chapter 9 of [12].

2. Satisficing Models

To facilitate our discussion, let us consider a simple MCDM problem:

Example 2.1. Mr. Davis is planning to make extra money on the coming Saturday. He has 14 hours available for work and leisure. And he would like to work at least five hours and spend no more than eight hours for leisure activities. Assume that he can earn $10 per hour and he can work as long as he wants at the same rate. In addition to making money, Mr. Davis has a life enjoyment goal which specifies one and five units of satisfaction respectively for each hour spent in work and leisure. Suppose Mr. Davis wants to make at least \$145 and to reach the total enjoyment of life of at least 50 units whenever possible.

Can Mr. Davis achieve his goals? And, if not, what would you suggest him to do?

According to Mr. Davis' story, let x_1 and x_2 be his time (measured in hours) allocated for work and leisure respectively. His two objective functions or goals can be specified by:

$$y_1 = f_1(x) = 10x_1 \tag{2.1}$$

$$y_2 = f_2(x) = x_1 + 5x_2 \tag{2.2}$$

the decision space X as a subset of R^2 is defined by

$$g_1(x) = x_1 + x_2 \leq 14 \tag{2.3}$$

$$g_2(x) = x_1 \geq 5 \tag{2.4}$$

$$g_3(x) = x_2 \leq 8 \tag{2.5}$$

and $x_1 \geq 0$, $x_2 \geq 0$.

His ideal goals are specified by

$$y_1 = f_1(x) = 10x_1 \geq 145, \text{ and} \tag{2.6}$$

$$y_2 = f_2(x) = x_1 + 5x_2 \geq 50 \tag{2.7}$$

From the example we see that Mr. Davis' goal functions are "more is better" and the goal values are adjustable. We shall continue on this example after we give a formal description of our models.

2.1. Goal Setting

Definition 2.1. Goal setting for satisficing models is defined as the procedure of identifying a satisficing set S such that, whenever the decision outcome is an element of S, the decision maker will be happy and satisfied and is assumed to have reached the optimal solution.

Let our MCDM problem be specified by the criteria $f = (f_1,\ldots,f_q)$ and the feasible set is $X = \{x \in R^n | g(x) \leqq 0\}$, where $g = (g_1,\ldots,g_m)$. There are two kinds of goal functions. In the first kind, goal values are adjustable and "more is better" (or "less is better"). For this kind of goal function, say $f_i(x)$, one may specify two critical values a_i and b_i (a_i for above and b_i for below), $b_i \leqq a_i$, such that whenever $f_i(x^0) < b_i$ (i.e., x^0 yields a value below the lower critical value for f_i), x^0 is unacceptable, and whenever $f_i(x^0) \geqq a_i$ (i.e., x^0 yields a value above the upper critical value for f_i), x^0 is acceptable with respect to f_i (although some other criteria may turn x^0 down). Note that since f_i is an adjustable criterion, its corresponding critical values $\{b_i,a_i\}$ may be subject to change as time and events change. For criteria of this kind we may add

$$f_i(x) \geqq b_i \tag{2.8}$$

to the constraints, so defining a new feasible set X; and

$$f_i(x) \geqq a_i \tag{2.9}$$

will be used to define the satisficing set S.

Observe that b_i and a_i may be interpreted as "the minimum limit" and "the aspiration level" for f_i, respectively. Also observe that whenever (2.9) is satisfied, (2.8) is satisfied. In satisficing solutions, (2.9) is important for modeling; in compromise solutions (2.8) is also important in the formulation. Note that when $a_i = b_i$, the two inequalities (2.8) and (2.9) become the same.

If the monotonicity is reversed, f_i exhibits "less is better." Then (2.8) and (2.9) can be replaced by (recall that $b_i \leqq a_i$)

$$f_i(x) \leqq b_i, \tag{2.10}$$

$$f_i(x) \leqq a_i, \tag{2.11}$$

where (2.10) defines the satisficing set S and (2.11) becomes a new constraint for X.

Note that in this case, b_i becomes the aspiration level and a_i becomes the maximum limit for f_i. Also note that by using $-f_i$ (instead of f_i), and $-b_i$ and $-a_i$ (instead of b_i, a_i), (2.10) and (2.11) reduce to (2.8) and (2.9).

The second kind of goal functions are those of which the target goal values are fairly fixed and not subject to much change; too much or too little is not acceptable. For this kind of goal function, say $f_j(x)$, one may again identify two values $\{b_j, a_j\}$ with $a_j \geqq b_j$ such that x is acceptable if and only if the following inequalities hold:

$$b_j \leqq f_j(x) \leqq a_j. \tag{2.12}$$

Note that whenever $a_j = b_j$, (2.12) reduces to an equality. Observe that b_j and a_j are two critical values of f_j so that whenever $f_j(x)$ is <u>below</u> b_j or <u>above</u> a_j, x is unacceptable.

In contrast to (2.12), there may be a criterion, say f_k, for which one may specify two parameters a_k, b_k, $a_k \geqq b_k$ such that x is acceptable iff

$$f_k(x) \leqq b_k \tag{2.13}$$

or

$$f_k(x) \geqq a_k. \tag{2.14}$$

For instance, $f_k(x)$ could be a position measurement from a potentially disastrous location (say, a nuclear power plant), for which one may specify that any acceptable site of a factory must satisfy

$$|f_k(x)| \geqq a_k,$$

where a_k is the minimum distance from the undesirable place. Using $b_k = -a_k$, the above inequality reduces to (2.13) and (2.14).

For future reference, let us summarize the above discussion in the following remark.

Remark 2.1. If the decision maker considers each criterion individually, one at a time, he can reach a satisficing set S which satisfies some or all of the inequalities of the types (2.8)–(2.14). Note that S so defined is a "rectangle" or the union of a number of "rectangles" [see (2.13) and (2.14)]. Also note that the satisficing set so defined does not reflect any consideration of tradeoffs among the criteria. When such consideration is important, one can define S by the union of

$$S_k = \{f | G_k(f) \geqq 0\}, \qquad k=1,\ldots,r, \tag{2.15}$$

where f is a point in S_k and G_k is a vector function which reflects the tradeoff over f (see Chapters 5 and 6 of [12] for further discussion). Note that when the union contains a single set (i.e., r=1), $S=S_1$ and S is defined by a system of inequalities.

Example 2.2. Let us continue on Mr. Davis' problem. From Example 2.1, we see that $S = \{f | f_1(x) - 145 \geq 0,\ f_2(x) - 50 \geq 0\}$, specifies Mr. Davis' satisficing set.

Example 2.3. Let us consider a problem of selecting an athlete for a basketball team. Assume quickness, f_1, and accuracy of shooting, f_2, of the player are essential. Let both f_1 and f_2 be indexed from 0 to 10, where the higher the index, the better the player. Assume that a player, x, will be selected with satisfaction if $f_1(x) \geq 9$ or $f_2(x) \geq 9$, or $f_1(x) + f_2(x) \geq 15$. Now set $G_1(f) = f_1(x) - 9$, $G_2(f) = f_2(x) - 9$, and $G_3(f) = f_1(x) + f_2(x) - 15$. We see that each $S_i = \{f | G_i(f) \geq 0\}$, i=1,2,3, specifies a satisficing set and $S = S_1 \cup S_2 \cup S_3$ is not convex.

2.2. Satisficing Model and Its Interactive Methods

Once S, the satisficing set, has been determined, finding a satisficing solution x^0, such that $f(x^0) \in S$, is a mathematical programming problem. Recall from Section 2.1 that when each criterion is considered individually the satisficing set S will be defined by inequalities of the types of (2.8)-(2.14); and when the criteria are jointly considered, then S may be defined by a form such as (2.15). As (2.15) is the most general form, we shall assume that S is defined by the union of S_k, k=1,...,r, with each S_k defined as in (2.15).

Since $S = \bigcup_{k=1}^{r} S_k$, $Y \cap S \neq \emptyset$ iff there is an S_k such that $Y \cap S_k \neq \emptyset$. As we need only one point of $Y \cap S$ when it is nonempty, we could verify individually if $S_k \cap Y \neq \emptyset$, $k=1,\ldots,r$, and find one point from the nonempty intersections.

Note that to verify whether or not $S_k \cap Y$ is empty and to find a point in $S_k \cap Y$, if it is nonempty, we can use the following mathematical program.

Rewrite each set in (2.15) as

$$S_k = \{f(x) \mid G_{kj}(f(x)) \geqq 0,\ j=1,\ldots,J_k\}, \tag{2.16}$$

where G_{kj}, $j=1,\ldots,J_k$, are components of G_k, and J_k is the number of components of G_k. Recall that all of the goal constraints (2.8)-(2.14), after properly rearranging the signs and terms, can be written as (2.16).

<u>Program 2.1.</u>

$$V_k = \min \sum_{j=1}^{J_k} d_j, \tag{2.17}$$

$$\text{s.t. } G_{kj}(f(x)) + d_j \geqq 0, \quad j=1,\ldots,J_k \tag{2.18}$$

$$x \in X, \quad d_j \geqq 0 \tag{2.19}$$

$$j=1,\ldots,J_k.$$

It is readily verified that $S_k \cap Y \neq \emptyset$ iff $V_k = 0$ in (2.17). We thus have the following theorem.

Theorem 2.1. Let $S = \bigcup_{k=1}^{r} S_k$. Then a satisficing solution exists iff there is at least one $k \in \{1,\ldots,r\}$ such that Program 2.1 yields $V_k = 0$.

Remark 2.2. The objective function in (2.17) can be replaced by any strictly increasing function of d_j, $j=1,\ldots,J_k$. In particular, one may use $\sum_{j=1}^{J_k} d_j^p$, $p \geq 1$. Then the resulting V_k measures the minimum ℓ_p-distance (see Section 3 for a further discussion) between Y and S_k. From a computational point of view the form (2.17), in which p=1, has the advantage of simplicity. Especially when G_{kj}, f, and g are all linear, Program 2.1 can be reduced to a linear program (i.e., a linear goal program which will be discussed in Section 3).

Remark 2.3. In view of (2.16), $G_{kj}(f(x))$ may be regarded as a newly defined criterion or goal function so that the nonnegative orthant $\Lambda^{\geqq}$ with respect to the G_{kj}'s is the satisficing set. With respect to these newly defined goal functions, goal setting becomes very simple. Program 2.1 can then be used to verify the existence of a satisficing solution by finding it when it exists.

Remark 2.4. In a degenerate case, the satisficing levels may be reduced to a point and (2.16) becomes

$$S_k = \{f(x) | G_{kj}(f(x)) = 0, \ j=1,\ldots,J_k\}. \tag{2.20}$$

In this case, Program 2.1 can be replaced by the following:

Program 2.2.

$$V_2 = \min \sum_{j=1}^{J_k} |G_{kj}(f(x))|$$

$$\text{s.t.} \quad x \in X.$$

Then the goal, 0, is obtained iff $V_2 = 0$. We thus have the following theorem.

Theorem 2.2. If S_k is defined as in (2.20), then $S_k \cap Y \neq \emptyset$ iff Program 2.2 yields $V_2 = 0$.

Remark 2.5. Comparing Program 2.1 with Program 2.2, we note that Program 2.2 has a simpler form of the objective function and may in general be easier to solve computationally than Program 2.1. However, Program 2.2 is not valid for a general satisficing set defined as in (2.16).

Remark 2.6. Observe that when one satisficing solution is found the decision problem is solved, at least temporarily. Otherwise the decision maker can activate either a positive problem solving or negative problem avoidance to restructure the problem. The former will enable careful restudy and restructuring of the problem so as to find a solution, perhaps a new one, which lies in the satisficing set. The latter will try to reduce the aspiration levels or play down the importance of making a good decision, thus, "lowering" the satisficing set to have a nonempty intersection with Y. While the psychological attitude in problem solving and in problem avoidance may be different (see

Chapter 9 of [12] for further discussion), the final consequence is the same. That is, eventually, the newly structured problem enables the decision maker to have $S \cap Y \neq \emptyset$. Here we will not discuss the restructuring of a problem to obtain a satisficing solution. Instead we will examine helpful information to aid the decision maker in restructuring his satisficing set S and feasible set Y. The following provides some helpful information.

(i) Depending on the formation of S, if Program 2.1 is used to identify a satisficing solution, one can produce x^k, $k = 1,\ldots,r$, which solves Program 2.1 with S_k as the satisficing set. One then can compute $f(x^k)$, $G_k(f(x^k))$, the distance from $f(x^k)$ to S_k, and tradeoff among the $G_{kj}(f(x))$, $j=1,\ldots,J_k$, at x^k. All of these can be helpful for the decision maker to reframe S_k. If Program 2.2 is used, 0 is the target for $G_k(f(x))$ to attain. The solution x^k will thus give a minimum distance from $G_k(f(x))$ to 0. Again, information on $f(x^k)$, $G_k(f(x^k))$ and the minimum distance from $G_k(f(x^k))$ to 0, $k=1,\ldots,r$, and the tradeoff among $G_{kj}(f(x^k))$, $j=1,\ldots,J_k$, at x^k can prove valuable. This information can be used by the decision maker to realize what he can attain and how far away those attainable points are from his satisficing set.

(ii) Suppose that it has been revealed or established that "more is better" for each criterion. Then the revealed preference $\{\succ\}$ contains at least the Pareto preference (see [12]). One can then generate: (A) all or some representatives of the N-points, the corresponding f(x), and optimal weights $\Lambda^*(x)$ for the N-points; (B) the value $f_k^* = \max\{f_k(x) | x \in X\}$, $k=1,\ldots,q$, which gives the best value of $f_k(x)$ over X. This information can help the decision maker to restructure his S.

Relaxation of the target goal with reference to the ideal point $(f_1^*,\ldots,f_q^*)$ sequentially and iteratively with the decision maker, until a satisficing solution is obtained, is certainly an art. It is especially useful when X and f are fairly fixed and not subject to change.

(iii) If possible (for instance, in bicriteria cases), displaying Y and S, even if only partially, can help the decision maker conceptualize where Y and S are, to make it easier to restructure the problem.

We summarize the concept of satisficing solutions and the procedure of using it to obtain a final satisficing solution in Figure 2.1.

Observe that Box (1) is clear. Box (2) is described in Sections 2.1 and 2.2. Boxes (5), (7) and (9) are clear for a conclusion of decision analysis. Note that related information includes those points in $S \cap Y$ and the corresponding points in X. For boxes (6) and (8), one can refer to Remark 2.6 and to Chapter 9 of [12] for further discussion.

One must keep in mind that with X, f(x) and S, one can generate as much information as one wishes, just as one can generate as many statistics from sample values as he wants. However, relevant information (e.g., useful statistics) which can positively affect the problem solving may not be too much. Irrelevant information can become a burden for decision analysis and decision making.

Remark 2.7. Everyone is subject to suggestions, including self-suggestion, to change his perception of decision outcomes and satisficing sets. As each individual is unique, there is no universal way that is most effective for every individual for changing the satisficing set

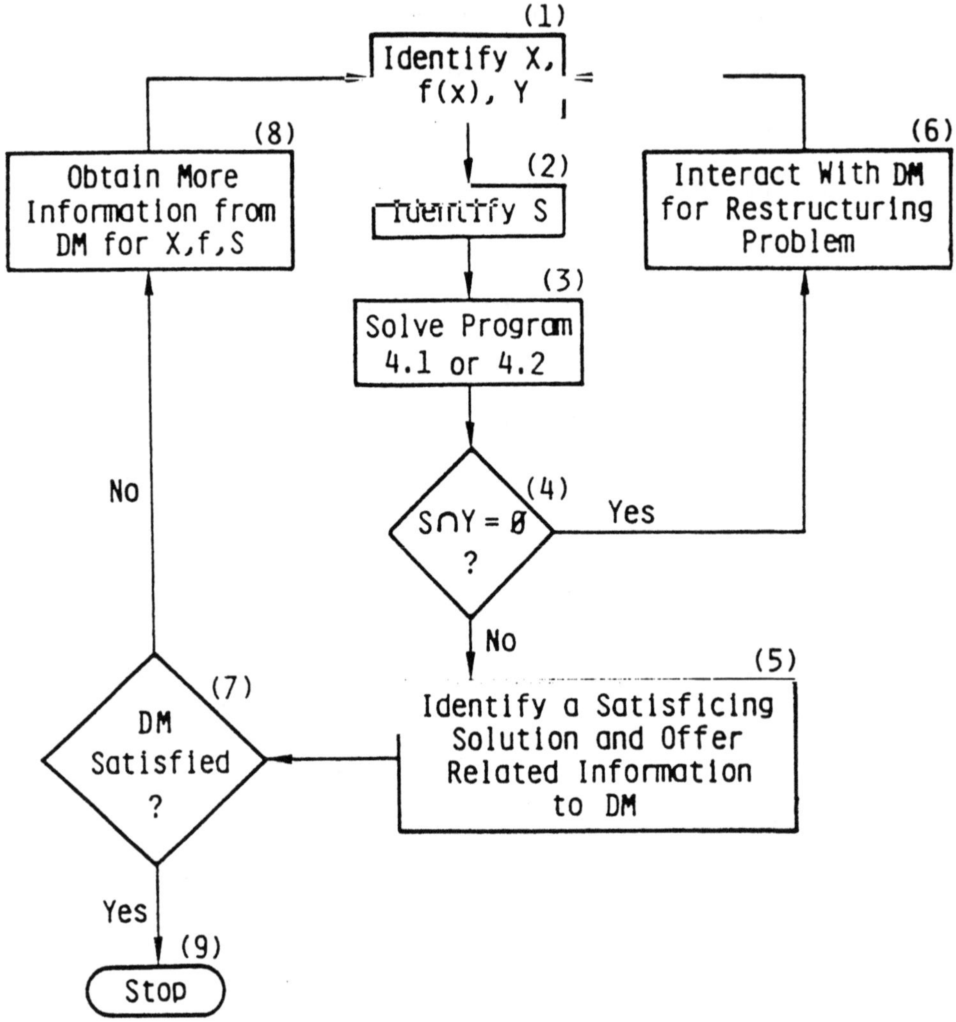
(1)
Identify X, f(x), Y
(2)
Identify S
(3)
Solve Program 4.1 or 4.2
(4)
S∩Y = Ø ?
Yes
No
(5)
Identify a Satisficing Solution and Offer Related Information to DM
(6)
Interact With DM for Restructuring Problem
(7)
DM Satisfied ?
No
Yes
(8)
Obtain More Information from DM for X,f,S
(9)
Stop

Figure 2.1

and obtaining a final solution. Changing or relaxing the aspiration levels of the criteria one at a time, two at a time, or all simultaneously, will depend on the individual. A good decision aid should be able to supply relevant information to facilitate proper change of the satisficing set in order to get a good final solution.

Now we are ready to handle Mr. Davis' problem.

Example 2.4. For illustration purpose, we draw the decision space X and the outcome space Y of Mr. Davis' problem in Figure 2.2 and Figure 2.3, respectively.

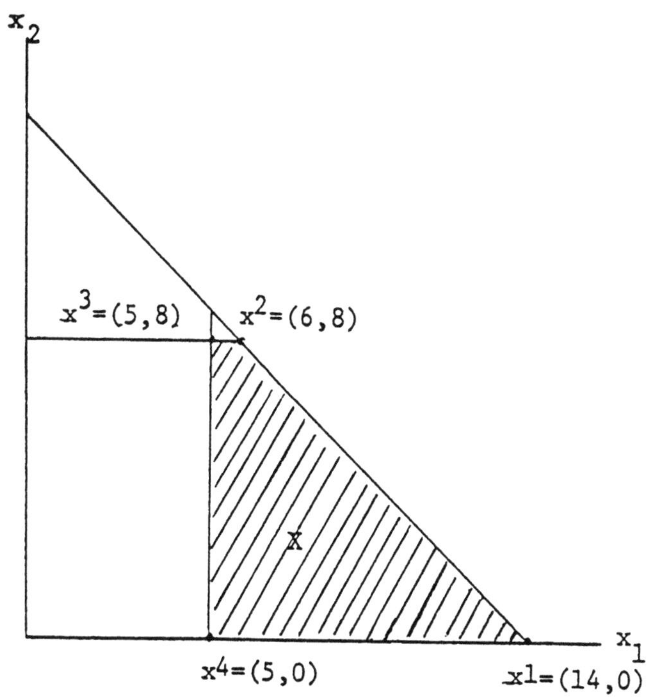

Figure 2.2 Decision Space X

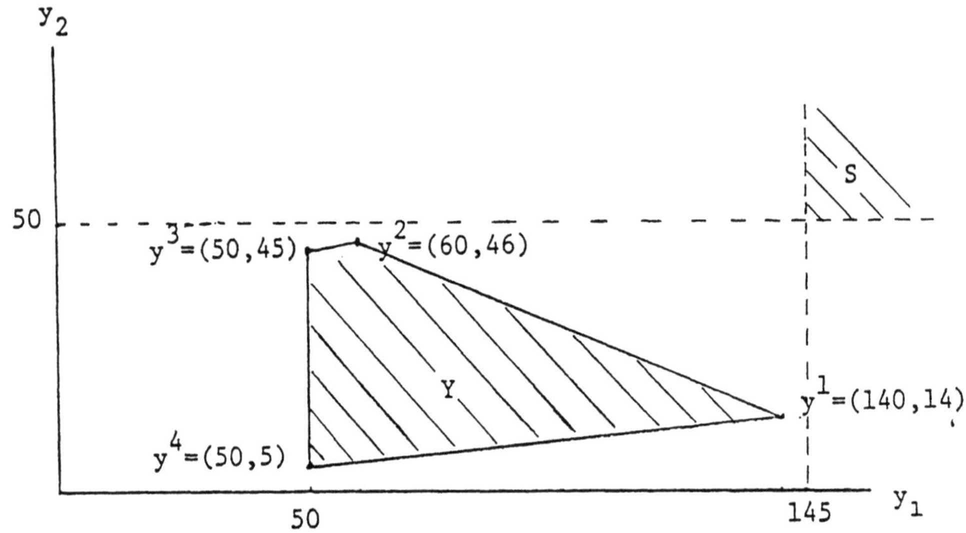

Figure 2.3 Outcome Space Y

In Example 2.2, we have known that Mr. Davis' initial satisficing set is $S = \{f(x) | 10x_1 - 145 \geq 0, x_1 + 5x_2 - 50 \geq 0\}$. Then using Program 2.1, we solve the following program:

$$
\begin{aligned}
V_1 = \min\ & d_1 + d_2 \\
\text{s.t. } & 10x_1 + d_1 \geq 145 \\
& x_1 + 5x_2 + d_2 \geq 50 \\
& x \in X \\
& d_1,\ d_2 \geq 0,
\end{aligned}
$$

where X defined in (2.3)-(2.5).

Observe from Figure 2.3, $S \cap Y = \emptyset$ as the solution to the above program $V_1 = 41 > 0$ indicated. Thus, Mr. Davis cannot reach his goals. We may suggest that he restructure X, f, and/or S to find a new satisficing solution. Now suppose that Mr. Davis restructure his satisficing set S by lowering down the aspiration levels, said $S' = \{f(x) | 10x_1 - 135 \geq 0,\ x_1 + 5x_2 - 16 \geq 0\}$. We see that the program becomes

$$V_1' = \min d_1' + d_2'$$
$$\text{s.t. } 10x_1 + d_1' \geq 135$$
$$x_1 + 5x_2 + d_2' \geq 16$$
$$x \in X$$
$$d_1', d_2' \geq 0,$$

which has optimal solution $V_1' = 0$. that means $S' \cap Y \neq \emptyset$ and there is a satisficing solution $y = (135, 16)$.

3. Compromise Models

In this section, mathematical formulation of goal seeking behavior in terms of a distance function will be described. Because it is simple to understand and compute, the concept has much general appeal to many scholars. We shall start with the most simple form and then gradually extend the concept to more complex cases. In Section 3.1, the basic concept and preference structure of compromise models are introduced. Section 3.2 is devoted to computational aspects including goal programming and maximin programming and Section 3.3 to a discussion on dynamic adjustment of ideal points for a final optimal solution.

3.1. Regret Functions and Compromise Solutions

Let the satisficing set S of Section 2.2 be a set of only one point y^*. That is, y^* is a unique target. Typically, when each criterion f_i,

$i=1,\ldots,q$, is characterized by "more is better," one can set $y^* = (y_1^*,\ldots,y_q^*)$ where $y_i^* = \sup\{f_i(x)|x \in X\}$. In this case, y^* is called an <u>ideal (or utopia) point</u> because usually y^* is not attainable. Note that, individually, the y_i^* may be attainable. But to find a point y^* which can simultaneously maximize each f_i, $i=1,\ldots,q$, is usually very difficult.

<u>Example 3.1</u>. To find Mr. Davis' ideal point, we solve the following two linear programs (recall Example 2.1).

For y_1^*, we maximize $y_1 = f_1(x) = 10x_1$ subject to $x \in X$ defined in (2.3)-(2.5). The solution is $x_1^* = 14$, $x_2^* = 0$. Thus, $y_1^* = f_1(x^*) = 140$. Similarly, for y_2^*, we maximize $y_2 = f_2(x) = x_1 + 5x_2$ subject to $\in X$ defined in (2.3)-(2.5). The solution is $x_1^* = 6$, $x_2^* = 8$. Thus, $y_2^* = f_2(x^*) = 46$. Therefore, Davis' ideal point is $y^* = (140, 46)$, which is specified in Figure 3.1.

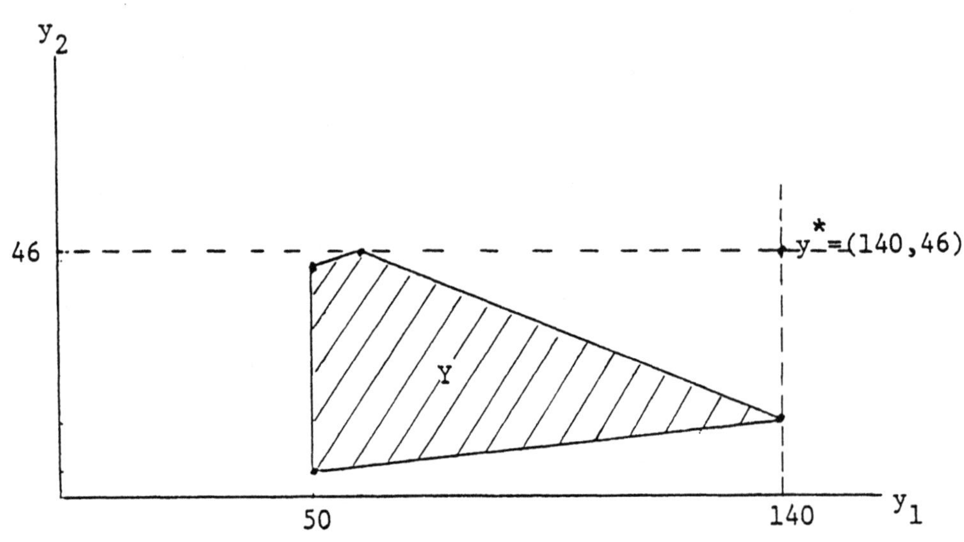

<u>Figure 3.1</u>

In group decision problems, if each criterion represents a player's payoff, then y*, if obtainable, would make each player happy because it simultaneously maximizes each player's payoff. Even if he is the dictator, he cannot do better than y* for himself. As y* is usually not attainable, to dissolve the group conflict, a compromise is needed if no other alternative is available. This offers a natural explanation of why the solution to be introduced is called a <u>compromise solution</u>. (See [10].) The reader can extend this explanation easily to multiple criteria problems.

Now, given $y \in Y$, the <u>regret</u> of using y instead of obtaining the <u>ideal</u> point y* may be approximated by the distance between y and y*. Thus, we define the (group) regret of using y by

$$r(y) = \|y - y^*\|, \tag{3.1}$$

where $\|y - y^*\|$ is the distance from y to y* according to some specified norm. Typically, the ℓ_p-norm will be used in our discussion because of its ease in understanding, unless otherwise specified. To make this more specific, define

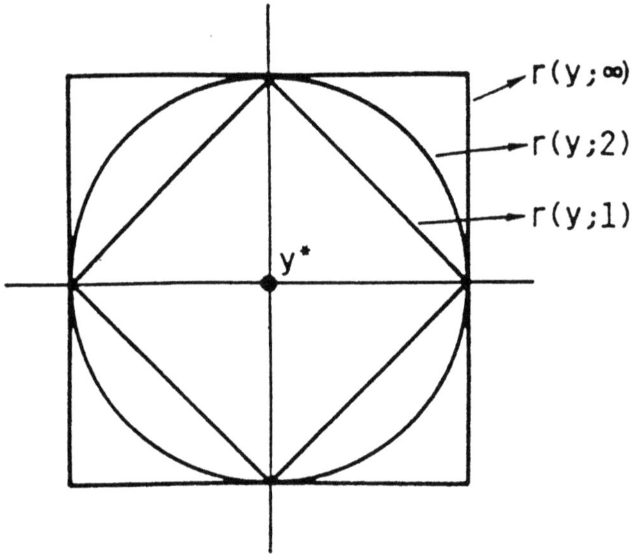

Figure 3.2

for $p \geqq 1$

$$r(y;\ p) = \| y - y^* \|_p = [\sum_i |y_i - y_i^*|^p]^{1/p} \tag{3.2}$$

and

$$r(y;\ \infty) = \max\{|y_i - y_i^*|,\ i = 1,\ldots,q\} \tag{3.3}$$

Then r(y;p) is a measurement of regret from y to y* according to the ℓ_p-norm. Note that the isovalued curves of r(y; p) centered at y* have some special forms as in Figure 3.2 for the case q = 2. They change shape from a "square diamond" for p = 1, to a circle for p = 2 and to a square for $p = \infty$.

Definition 3.1. The compromise solution with respect to the ℓ_p-norm is $y^p \in Y$, which minimizes r(y; p) over Y; or is $x^p \in X$, which minimizes r(f(x); p) over X.

Example 3.2. Using simple isovalued curves of Figure 3.2, we find compromises solutions $y^1 = (140, 14)$, $y^2 = (128, 18)$, and $y^\infty = (117.5, 23)$ for Mr. Davis in Figure 3.3., where y^1 minimizes $r(y;1)$, y^2 minimizes $r(y;2)$, and y^∞ minimizes $r(y;\infty)$. The mathematical program for finding compromise solutions will be discussed in Section 3.2.

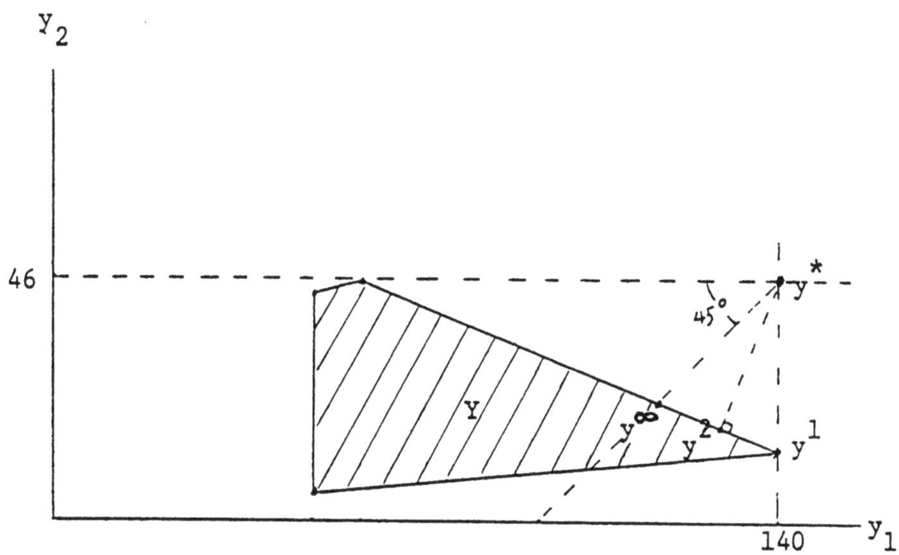

Figure 3.3

Remark 3.1. In group decision problems, $r(y; p)$ may be interpreted as group regret and the compromise solution y^p is the one which minimizes the group regret in order to maintain a "cooperative group spirit." As the parameter p varies, the solution y^p can change. We shall list the impact of p on the solutions in the Appendix.

Note that $r(y; p)$ treats each $|y_i^* - y_i|$ as having the same importance in forming the group regret. In multiple criteria problems, if the criteria have different degrees of importance, then a weight vector

$w = (w_1, w_2, \ldots, w_q)$, $w \geq 0$ may be assigned to signal the different degrees of importance. In this case, we define

$$\begin{aligned} r(y;\ p,\ w) &= \|y - y^*\|_{p,w} \\ &= [\sum_i w_i^p |y_i - y_i^*|^p]^{1/p} \\ &= [\sum_i |w_i y_i - w_i\ y_i^*|^p]^{1/p} \end{aligned} \tag{3.4}$$

The concept of compromise solutions can then naturally be extended to include the regret function of (3.4).

Definition 3.2. The compromise solution with respect to $r(y;\ p,w)$ of (3.4) is $y^{pw} \in Y$, which minimizes $r(y;\ p,\ w)$ over Y; or is $x^{pw} \in X$, which minimizes $r(f(x);\ p,\ w)$ over X.

Remark 3.2. Observe that the weight vector w as in (3.4) has the effect of changing the scale of each criterion. Once the scale is adjusted (i.e., by $w_i y_i = w_i f_i(x)$ for all $i = 1, \ldots, q$), the regret function is reduced to that of equal weight. Thus, in studying the properties of compromise solutions, without loss of generality, one may focus on the equal weight case. We shall so assume from now on, unless specified otherwise.

Remark 3.3. Observe that the compromise solution is not scale independent. Scale independence, an important criterion in group decision problems, can prevent players from artificially changing the scale to obtain a better arbitration for themselves. Note that when the scale of $f_i(x)$ or y_i is changed and the weight of importance of each criterion is changed, so is the compromise solution.

Definition 3.3. Given $r(y; p)$, define the preference using $r(y; p)$ by $y^1 \succ y^2$ iff $r(y^1; p) < r(y^2; p)$.

Theorem 3.1. The preference using $r(y; p)$ is a weak order and, for any y^1 and y^2, $y^1 \sim y^2$ if and only if $r(y^1; p) = r(y^2; p)$.

Remark 3.4. Theorem 3.1 reveals that the concept of compromise solutions implicitly has assumed a very strong preference assumption. There are no ambiguity in preference comparison. For any two points y^1 and y^2, either y^1 is better or worse or equivalent to y^2. Exactly one of the three cases must happen. There are no such things as "indefinite" in the preference of compromise solutions. Before we finally use it, precaution is needed to verify that the assumption is fulfilled; otherwise, a misleading solution may result.

3.2. Computing Compromise Solutions

To find a compromise solution, one needs to know the target point, either the ideal point or not, and the parameter p for the regret function $r(f(x); p)$. When the target point is unknown, one may start with the ideal point as an approximation of the target. Hopefully, by an interactive process, one can gradually locate the target point and the compromise solution. When the ideal point is the target point, computing for the compromise solution is not very complicated, at least theoretically. This will be discussed in Section 3.2.1. For general target points the computation can be slightly more complex. We shall discuss this in Section 3.2.2.

3.2.1. The Ideal Point as the Target Point

To find a compromise solution when the ideal point is the target point, we have to solve q + 1 mathematical programming problems. The first q problems are to find the utopia or ideal point, $y^* = (y_1^*, \ldots, y_q^*)$ where $y_i^* = \max\{f_i(x) | x \in X\}$. The last one is to find the compromise solution y^p. Suppose that X contains only countable points. We will have q + 1 integer programming problems. Otherwise, if X is a region, we will have q + 1 nonlinear programming problems.

Suppose X and $f_i(x)$, i=1,...,q, have some special structure. More efficient computational techniques are available. For instance, if X is a convex set and each $f_i(x)$ is concave, then we have first q concave programming and then a convex programming (because $r(f(x); p)$ is convex under the assumptions). If X is a polyhedron defined by a system of linear inequalities and each $f_i(x)$ is linear, then the ideal points y^* can be found by q simple linear programming problems. Furthermore, the compromise solutions of y^1 and y^∞ can be found by a linear programming problem (the other compromise solution y^p, $1 < p < \infty$ can be found by convex programming). That y^1 can be found by linear programming is obvious. That y^∞ can be found by linear programming follows directly from the following well-known result.

<u>Lemma 3.1.</u> x^0 solves

<u>Problem A:</u> $\min_x \max_i \{y_i^* - f_i(x) | i = 1,\ldots,q, x \in X\}$

if and only if x^0 and $v^0 = \max_i\{y_i^* - f_i(x^0)\}$ solve

<u>Problem B:</u> $\min\{v | v \geq y_i^* - f_i(x), i = 1,\ldots,q, x \in X\}$.

Example 3.3. Let us use mathematical programs to find the compromise solutions y^1, y^2, and y^∞ for Mr. Davis when the target point is the ideal point.

From Example 3.1, we know that the ideal point $y^* = (140, 46)$. Thus,

(i) for the y^1 compromise solution, we solve

$$\min \{[140 - 10x_1] + [46 - (x_1 + 5x_2)]\},\ x = (x_1, x_2) \in X.$$

We obtain $x_1^* = 14$, $x_2^* = 0$, and $y^1 = (140, 14)$.

(ii) for the y^2 compromise solution, we solve

$$\min \{[140 - 10x_1]^2 + [46 - (x_1 + 5x_2)]^2\},\ x = (x_1, x_2) \in X.$$

The solution is $x_1^* = 12.8$, $x_2^* = 1.04$ and $y^2 = (128, 18)$.

(iii) For the y^∞ compromise solution, we solve

$$\min \max [140 - 10x_1,\ 46 - (x_1 + 5x_2)],\ x = (x_1, x_2) \in X.$$

According to Lemma 3.1, the above problem is simplified as follows:

$$\begin{aligned} &\min v \\ &\text{s.t. } v \geqq 140 - 10x_1 \\ &\qquad v \geqq 46 - (x_1 + 5x_2) \\ &\qquad x \in X. \end{aligned}$$

We obtain

$$x_1^* = 11.75,\ x_2^* = 2.25, \text{ and } y^\infty = (117.5, 23).$$

The following table is a summary of compromise solutions, when $p = 1, 2$, and ∞, with respect to the ideal point $y^* = (140, 46)$.

Compromise Solutions

	(x_1,x_2)	(y_1,y_2)
$p = 1$	(14,0)	$y^1 = (140,14)$
$p = 2$	(12.8, 1.04)	$y^2 = (128,18)$
$p = \infty$	(11.75, 2.25)	$y^\infty = (117.5,23)$

Observe that the above solutions are identical to those of the graphical solutions in Example 3.2. Also observe Figure 3.3 in terms of properties 5.8 and 5.11 in the Appendix, as a function of p, y^p moves continuously and monotonically from y^1 to y^∞.

3.3.2. General Target Points and Goal Programming

Suppose that the target point for compromise solutions is known but is not the ideal point. Then each f_i may not be the type of "more is better." For instance, if the target for f_i is 10 and $f_i[X] = [8,20]$, then $f_i = 15$ is attainable but not as good as $f_i = 10$. Thus the preference is not monotone over f. However, moving as close to the target as possible is still the same. Note the resulting compromise solution for the general target point may not be Pareto optimal over Y. Indeed Pareto optimality does not make sense here as preference is no longer monotone over f. To regain the monotonicity, one needs to redefine the criteria. For instance, $f_i = \| f_i - y_i^* \|$, where y_i^* is the target of f_i, for smaller is better and $\bar{f}_i = M - \| f_i - y_i^* \|$, with M sufficiently large, for larger is better.

The computation for the compromise solution with general target point can be slightly more complicated. This is due to the possibility that $y_i^* \geq y_i$ for all $y \in Y$ and $i = 1,\ldots,q$ may no longer hold. The following method of changing variables can be of help.

Lemma 3.2. Let

$$d_i^+ = \begin{cases} y_i - y_i^* & \text{if } y_i > y_i^* \\ 0 & \text{otherwise;} \end{cases}$$

$$d_i^- = \begin{cases} y_i^* - y_i & \text{if } y_i < y_i^* \\ 0 & \text{otherwise.} \end{cases}$$

Then

(i) $y_i^* - y_i = d_i^- - d_i^+$

(ii) $|y_i^* - y_i| = d_i^- + d_i^+$

(iii) $d_i^+, d_i^- \geq 0$

Note that d_i^+ is the value of y_i exceeding y_i^* (like a surplus variable) and d_i^- is the value of y_i below y_i^* (like a slack variable). The proof of the lemma is straightforward.

Lemma 3.3. The definition of d_i^+ and d_i^- in Lemma 3.2 is satisfied if and only if d_i^+ and d_i^- are nonnegative and satisfy:

$$y_i^* - y_i = d_i^- - d_i^+ \tag{3.5}$$

$$d_i^+ \cdot d_i^- = 0 \tag{3.6}$$

Now given a target point y^* the compromise solution with ℓ_p-distance can be found by solving

Program 3.1.

$$\min \sum_i |y_i^* - f_i(x)|^p = \min [r(f(x); p)]^p \tag{3.7}$$

$$\text{s.t. } g(x) \leq 0 \tag{3.8}$$

Note that since $[r(f(x); p)]^p$ is a strictly increasing function of $r(f(x); p)$, x^0 minimizes $[r(f(x); p)]^p$ if and only if it minimizes $r(f(x); p)$. The form of (3.7) makes it easier to find the compromise solution.

We now can apply Lemma 3.2 and 3.3 to remove the sign of the absolute values in (3.7) and get

Program 3.2.

$$\min \sum_{i=1}^{q} (d_i^- + d_i^+)^p \tag{3.9}$$

$$\text{s.t. } y_i^* - f_i(x) = d_i^- - d_i^+ \tag{3.10}$$

$$d_i^+ , d_i^- \geqq 0 \tag{3.11}$$

$$g(x) \leqq 0 \tag{3.12}$$

where $i=1,2,\ldots,q$ in (3.10)-(3.12).

Theorem 3.2. Given a target point y^*, x^0 is the compromise solution with respect to the ℓ_p-distance if and only if x^0 solves Program 3.2.

Remark 3.5. (i) Suppose that all $f_i(x)$, $i = 1,\ldots,q$, and $g(x)$ are linear. Then Program 3.2 for ℓ_1-compromise solutions reduces to a linear program typically known as (linear) goal programming. By adding weights to or imposing lexicographical ordering on the criteria one can generalize the concept discussed here to a variety of goal programming formats. One notes that goal programming is a special class in the domain of compromise solutions. The major advantages of goal programming and ℓ_1-compromise solutions are that they are easily understood and that they can be easily computed by linear programs. In applications, one should be aware of its endowed properties as discussed in the Appendix to avoid misused or misled conclusions.

(ii) When all $f_i(x)$, $i=1,\ldots,q$, and $g(x)$ are linear, Program 3.2 for ℓ_2-compromise solutions becomes a quadratic program. The program can be solved without major difficulty.

(iii) When all $f_i(x)$, $i=1,\ldots,q$, and $g(x)$ are linear, Program 3.2 for ℓ_∞-compromise solutions becomes a linear program by applying Lemma 3.1. This greatly simplifies the computations. Again, one should understand the properties endowed by the ℓ_∞-compromise solution as discussed in the Appendix to avoid misuse.

Example 3.4. Suppose instead of the ideal point $y^* = (140,46)$, Mr. Davis sets (130,46) as his target point. We can find the compromise solutions by using Program 3.2 as follows:

(i) For the y^1-compromise solution, we solve

$$\begin{aligned}
\min\ & (d_1^- + d_1^+) + (d_2^- + d_2^+) \\
\text{s.t.}\ & 130 - 10x_1 = d_1^- - d_1^+ \\
& 46 - (x_1 + 5x_2) = d_2^- - d_2^+ \\
& x_1 + x_2 \leq 14 \\
& x_1 \geq 5 \\
& x_2 \leq 8 \\
& x_1,\ x_2 \geq 0,\ d_1^-,\ d_1^+,\ d_2^-,\ d_2^+ \geq 0.
\end{aligned}$$

The solution is $d_1^- = d_1^+ = 0$, $d_2^- = 28.5$, $d_2^+ = 0$, $x_1 = 13$, and $x_2 = 0.9$. Thus, the y^1-compromise solution is $y^1 = (130,17.5)$.

(ii) For the y^2-compromise solution, we solve

$$\min\ (d_1^- + d_1^+)^2 + (d_2^- + d_2^+)^2$$

$$\text{s.t.} \quad 130 - 10x_1 = d_1^- - d_1^+$$

$$46 - (x_1 + 5x_2) = d_2^- - d_2^+$$

$$x_1 + x_2 \leqq 14$$

$$x_1 \geqq 5$$

$$x_2 \leqq 8$$

$$x_1,\ x_2 \geqq 0,\ d_1^-,\ d_1^+,\ d_2^-,\ d_2^+ \geqq 0.$$

We get $d_1^- = 11$, $d_1^+ = 0$, $d_2^- = 23.5$, $d_2^+ = 0$, $x_1 = 11.9$, and $x_2 = 2.12$. Therefore, the y^2-compromise solution is $y^2 = (119, 22.5)$.

(iii) For the y^∞-compromise solution, we have

$$\min\ \max[d_1^- + d_1^+,\ \ d_2^- + d_2^+]$$

$$\text{s.t.} \quad 130 - 10x_1 = d_1^- - d_1^+$$

$$46 - (x_1 + 5x_2) = d_2^- - d_2^+$$

$$x_1 + x_2 \leqq 14$$

$$x_1 \geqq 5$$

$$x_2 \leqq 8$$

$$x_1,\ x_2 \geqq 0,\ d_1^-,\ d_1^+,\ d_2^-,\ d_2^+ \geqq 0.$$

Using Lemma 3.1, we simplify the above program as follows:

$$\begin{aligned}
&\min\ v \\
&\text{s.t.}\quad v \geqq d_1^- + d_1^+ \\
&\qquad v \geqq d_2^- + d_2^+ \\
&\qquad 130 - 10x_1 = d_1^- - d_1^+ \\
&\qquad 46 - (x_1 + 5x_2) = d_2^- - d_2^+ \\
&\qquad\qquad x_1 + x_2 \leqq 14 \\
&\qquad\qquad x_1 \geqq 5 \\
&\qquad\qquad x_2 \leqq 8 \\
&x_1,\ x_2,\ d_1^-,\ d_1^+,\ d_2^-,\ d_2^+ \geqq 0.
\end{aligned}$$

Solving the program, we get $d_1^- = 20$, $d_1^+ = 0$, $d_2^- = 20$, $d_2^+ = 0$, $x_1 = 11$, $x_2 = 3$, and $v = 20$. The y^∞-compromise solution is $y^\infty = (110,26)$.

We summary the y^p-compromise solutions, when $p = 1$, 2, and ∞, with respect to the target point $y^* = (130, 46)$ as the following table.

Compromise Solutions

	(x_1,x_2)	(y_1,y_2)
p=1	(13, 0.9)	$y^1 = (130,17.5)$
p=2	(11.9, 2.12)	$y^2 = (119,22.5)$
p=∞	(11, 3)	$y^\infty = (110,26)$

3.3. Interactive Methods of Compromise Models

For nontrivial decision problems, an optimal decision usually cannot be reached by applying the compromise solution only once. The following interactive method, according to Figure 3.4, may be helpful.

Box (1) Identifying X, f and y* is an art that requires careful observation and conversation with the decision maker. (See Chapter 9 of

[12] for further discussion). The target point y*, weight vector w and p for ℓ_p-distance may be more difficult to specify precisely. However, one can use the ideal point as the first step of approximation for y*, select some representative weights for w and let p = 1, 2, ∞ to begin.

Box (2) One can locate compromise solutions according to Program 3.2 for the specified values of y*, w and p.

Box (3) All relevant information obtained in Box (1) and (2) can be presented to the decision maker. These include X, f, Y, ideal points y_i^*, i=1,...,q, the optimal vector value which maximizes the ith criterion, the compromise solutions y^1, y^2, y^∞ with their weights and norms.

Box (4) and (6) If an optimal solution is reached in Box (3) and the decision maker is satisfied, the process is terminated at Box (6); otherwise, we go to Box (5) to obtain more information.

Box (5) Through conversation, we may obtain new information on X, f, y*, weight vector w and norm parameter p. Note that each of these five elements may change with time. To help the decision maker locate or change the target point, one may use the "one at a time method." That is, only one value of f_i or y_i, i=1,...,q, is to be changed and the rest remain unchanged. One may ask the decision maker: "Since no satisfactory solution is obtained with the current target point y*, would it be possible to decrease (or increase) the value of y_i^*? And, if so, by how much?" This kind of suggestive question may help the decision maker think hard and carefully of possible changes in the target points. The other method is called the "pairwise trade-off method." We first select two criteria, say f_i and f_j, and then ask: "To maintain the same degree of satisfaction for the target point, everything else

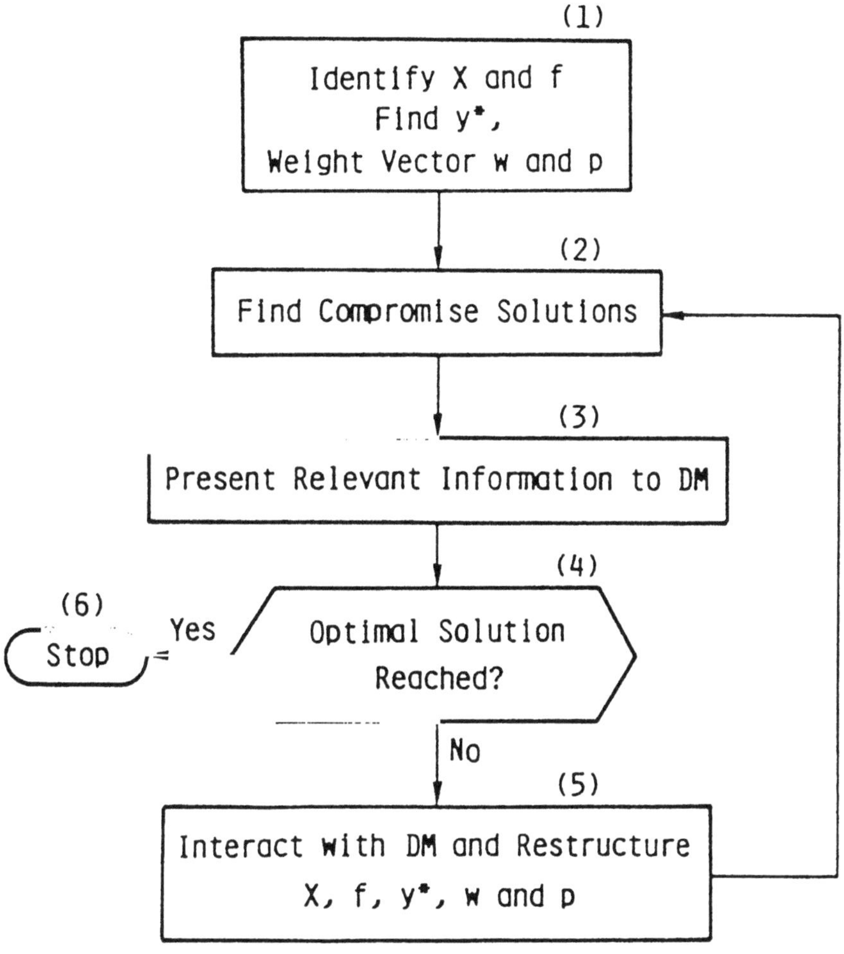
(1)
Identify X and f
Find y*,
Weight Vector w and p
(2)
Find Compromise Solutions
(3)
Present Relevant Information to DM
(4)
Optimal Solution
Reached?
(6)
Stop
Yes
No
(5)
Interact with DM and Restructure
X, f, y*, w and p

Figure 3.4

being equal, how many units of f_j must be increased in order to compensate a one-unit decrease in f_i? Again, this kind of suggestive question can help the decision maker think hard and carefully of the possible changes in the target points. Finally, we recall that the target point may be regarded as the satisficing set containing a single satisficing solution. Thus, by locating the satisficing set, we may locate the target point.

Remark 3.6. Comparing Section 2 with Section 3, we know that satisficing models are more general and flexible in applications, while compromise models are more specific and rigid. The former is loose while the latter makes sense only when strong assumptions are satisfied. However, with flexible interpretation and artful applications in interaction with the decision maker, the two methods can complement each other. After all, both concepts stem from the goal seeking assumption and try to minimize the distance from the criteria space to a set of satisficing or a target point. In a broader perspective, they are the same.

Example 3.5. Assume that Mr. Davis understands the above interactive method. Our interactive process with him to reach a final solution can be as follows:

Step 1. Identify X, f, y*, w, and p. The decision space X, is given by

$$g_1(x) = x_1 + x_2 \leqq 14$$

$$g_2(x) = x_1 \qquad \geqq 5$$

$$g_3(x) = \qquad x_2 \leqq 8$$

$$x_1, x_2 \geqq 0$$

and the objective function f_1 and f_2 are

$$y_1 = f_1(x) = 10x_1$$

$$y_2 = f_2(x) = x_1 + 5x_2$$

Through conversation with Mr. Davis, it is reasonable to assume that the ideal point is a good approximation for the target point. The ideal point (y_1^*, y_2^*) is (146,46) (see Example 3.1). Mr. Davis agrees that each criterion carries the same weight (i.e., $w_1 = w_2$). We use $p = 1,2,\infty$ to find the compromise solutions.

Step 2. Find Compromise Solutions. Refer to Example 3.2 or 3.3, we know that the compromise solutions are $y^1 = (140, 14)$, $y^2 = (128,18)$, $y^\infty = (117.5,23)$.

Step 3. Present Relevant Information to DM. All relevant information in Steps 1 and 2 is represented to Mr. Davis. Since we are in the two-dimension case, the graphic display in Figure 3.3 can greatly help Mr. Davis to visualize what he can and cannot reach.

Step 4. Optimal Solution Reached? Assume that Mr. Davis is not satisfied with and the compromise solutions presented to him in Step 3.

Step 5. Restructure New Model. Suppose Mr. Davis, through conversation with us, prefers to move the target point closer to outcome space Y by reducing \$10 from y_1^* and maintain the equal weights and p values.

Thus, the new target point is $y^* = (130,46)$ and the new compromise solutions as illustrated in Example 3.4 are $y^1 = (130, 31.5)$, $y^2 = (126,33)$, and $y^\infty = (118,34.5)$ can be located. Once more, suppose that Mr. Davis is still not satisfied with those new compromise solutions, and tells us that his importance weights over the criteria are different, say $w = (3/4, 1/4)$, and X, f, p, and y^* as in Example 3.4 remain unchanged. We can again recalculate y^1, y^2 and y^∞. The process will continue until Mr. Davis is satisfied with some solutions or quits.

4. Conclusions

Utilizing human goal-seeking behavior, we have described the main concepts of satisficing and compromise solutions. In a real world, although specifying the satisficing set or the target point is not trivial, we can always start with the ideal point and then interact with the decision maker as to gradually reach the final decision. For more details see Chapter 4 of [12]. There is a rich literature related to satisficing and compromise models. For instance, for goal programming see [1], [6] and those quoted therein; for different kinds of norms or penalty functions see [5], [7], and [8]; for restructuring ideal points see [13]; and for social science approach of compromise solutions see [2] and [3].

5. Appendix: General Properties of Compromise Solutions

Many solution concepts have been proposed in MCDM problems. However, few theoretical study of properties of those solutions have been thoroughly done. Understanding each property more precisely is very useful to apply the solutions into a real world without misleading. We briefly list the following properties of compromise solutions reported in [4] and [10] to help the reader know the compromise models better. For details, see Chapter 4 of [12].

For the purpose of simplicity, we make the following assumption throughout this section.

Assumption 5.1. (i) Y is an nonempty compact set. (ii) The single satisficing point y* is the ideal point, i.e., $y_i^* = \sup\{f_i(x) | x \in X\}$; (iii) The preference revealed for each criterion is characterized by "more is better."

Then we have the following properties:

Property 5.1 (Feasibility). The compromise solution y^p and $x^p \in f^{-1}(y^p)$ are obtainable or feasible.

Property 5.2 (Least Group Regret). If r(y;p) represents the group regret of using y instead of reaching y*, then y^p, by definition, minimizes the group regret.

Property 5.3 (No Dictatorship). Since each criterion is used to form r(y;p) to obtain y^p, there is no dictatorship in the process. That is, there is no sole j (dictator) such that y^p is completely determined by $f_j(x)$.

Property 5.4 (Pareto Optimality). For $1 \leqq p < \infty$, y^p is a nondominated point (or efficient point) with respect to Pareto preference. If y^∞ is unique, then y^∞ is also a nondominated point.

Property 5.5 (Uniqueness). For $1 < p < \infty$, y^p is unique if Y is $\Lambda^{\leqq}$ - convex (i.e., $Y + \Lambda^{\leqq}$ is a convex set). If Y is strictly convex, then y^1 and y^∞ are also unique.

<u>Property 5.6 (Symmetry)</u>. If Y is convex and closed with respect to cyclical notation, then for $1 < p \leq \infty$, $y_k^p = y_j^p$ for all k, $j=1,\ldots,q$. When $p = 1$, then there is at least one y^1 with $y_k^1 = y_j^1$ for all k, $j=1,\ldots,q$.

<u>Property 5.7 (Independence of Irrelevant alternatives)</u>. Let $Y^0 \subset Y$ be such that $\sup\{y_i | y \in Y^0\} = y_i^* = \sup\{y_i | y \in Y\}$, $i=1,\ldots,q$. If $y^p \in Y^0$, $1 \leq p \leq \infty$, then y^p is also the compromise solution with respect to $r(y;p)$ over Y^0.

<u>Property 5.8 (Continuity)</u>. If Y is compact and $\Lambda^{\geq}$-convex, then the compromise solution y^p is a continuous function of p for $1 < p < \infty$. In addition, if y^1 is unique, then y^p is continuous at $p = \infty$.

Let L be the line of equal regrets, i.e., $y_i^* - y_i = y_j^* - y_j$, $i \neq j$, i, $j=1,\ldots,q$, and $B_p = \{y \in R^q | r(y;p) = r(y^p,p)\}$.

Let v_p be the point of intersection of B_p with the positive half of L. Then, we define $d^p = y_i^* - v_i^p$, $i=1,\ldots,q$.

<u>Property 5.9 (Monotonicity in General)</u>. The function d^p is monotone nondecreasing for $1 \leq p \leq \infty$.

<u>Property 5.10 (Bounds)</u>. The following bounds hold:

(i) $$\sum_{i=1}^{q} v_i^{\infty} \leq \sum_{i=1}^{q} y_i^s \leq \sum_{i=1}^{q} y_i^1, \quad 1 \leq s \leq \infty$$

(ii) If y^{∞} lies on the line L of equal regrets, then

$$\sum_{i=1}^{q} y_i^{\infty} \leq \sum_{i=1}^{q} y_i^s \leq \sum_{i=1}^{q} y_i^1, \quad 1 \leq s \leq \infty.$$

<u>Property 5.11 (Monotonicity in Two Criteria)</u>. If $q=2$ and $1 \leqq s \leqq p \leqq \infty$, then

(i) $\sum_{i=1}^{q} y_i^p \leqq \sum_{i=1}^{q} y_i^s$;

(ii) $\max_i \{y_i^* - y_i^p\} \leqq \max_i \{y_i^* - y_i^s\}$;

(iii) In addition, if Y is compact and $\Lambda^{\leqq}$-convex, then y_1^p is a monotone function of p. If y_1^p is monotonically increasing, then y_2^p is monotonically decreasing, and vice versa.

References

1. Charnes, A., and Cooper, W. W., Management Models and Industrial Applications of Linear Programming, Vol. I, Wiley, New York (1961)

2. Davis, O. A., Degroot, M. H., and Hinich, M. J., "Social Preference Orderings and Majority Rule," Econometrica 40, 147-157 (1972).

3. Dinkelbach, W., Entscheidungs-Modelle, Walter de Gruyter, Berlin (1982).

4. Freimer, M., and Yu, P. L., "Some New Results on Compromise Solutions for Group Decision Problems," Management Science 22, 688-693 (1976).

5. Gearhart, W. B., "Compromise Solutions and Estimation of the Noninferior Set," J. Optim. Theory Appl. 28, 29-47 (1979).

6. Lee, S. M., Goal Programming for Decision Analysis, Auerbach, Philadelphia, Pennsylvania (1972).

7. Lindsay, P. H., and Norman, D. A., Human Information Processing, An Introduction to Psychology, Academic, New York (1972).

8. Pascoletti, A., and Serafini, P., "Scalarizing Vector Optimization Problems," J. Optim Theory Appl. 42, 499-524 (1984)

9. White, D. J., "Multiple Objective Programming and Penalty Functions," J. Optim. Theory and Appl. 43, 583-599 (1984).

10. Yu, P. L., "A Class of Solutions for Group Decision Problems," Management Science 19, 936-946 (1973).

11. Yu, P. L., "Behavior Bases and Habitual Domains of Human Decision/Behavior: An Integration of Psychology, Optimization Theory, and Common Wisdom," International Journal of Systems, Measurement and Decisions, 1, 39-62 (1981).

12. Yu, P. L. Multiple Criteria Decision Making: Concepts, Techniques and Extensions, Plenum, New York, New York, (1985).

13. Zeleny, M., "The Theory of the Displaced Ideal," in Multiple Criteria Decision Making: Kyoto 1975, M. Zeleny (ed.), Springer-Verlag, New York, 151-205, (1975).

The TRIMAP PACKAGE as a POWERFUL TEACHING TOOL

Joao Clímaco
Associate Professor
Department of Electrical Engineering
University of Coimbra
3000 COIMBRA PORTUGAL

SUMMARY

TRIMAP is an interactive tricriteria linear programming method based on a progressive and selective "learning" of the Pareto optimal solutions set [1]. The intention is not to converge into the "best" compromise solution but to help the decision maker to eliminate the sub-sets of solutions which are of no interest to him. This reduces the calculations and leads to a limited number of solutions for final selection. There are no irrevocable decisions along the process, being always admissible to go "backwards". The method is applicable to problems with three objective functions. Although being a limitation this allows the use of graphical means specially convenient for the "dialogue" with the decision maker (DM). In this paper we explore the teaching potentialities of TRIMAP.

The theoretical justifications of the used procedures are ommited. In essence these justifications may be found in other papers included in this publication, namely in S. Zionts [7]. For further details use the excellent book by R. Steuer [4]. Some examples shown in [7] are here used for helping the reader to "paralell" reading of these two papers.

NATO ASI Series, Vol. F56
Multiple Criteria Decision Making and Risk Analysis
Using Microcomputers. Edited by B. Karpak and S. Zionts

PART I

TRIMAP MAIN FEATURES

The planning of new units for electrical power production is a problem whose study explicitly requires several objective functions. We have carried out a study some years ago, using a linear model which considered three objective functions: the global cost, a second one related to the risk of supply and a third one related to environmental impacts. A multiobjective linear programming algorithm by Zeleny for calculating the set of the Pareto optimal extreme points was used [6]. However further experiments with more realistic formulations, leading to a much higher number of variables and constraints, confirmed the problems raised in these cases by the application of the method for generating all the Pareto optimal extreme points. Namely, the computational effort becomes impracticable when the Pareto optimal extreme points grow beyond certain limits; and even when it is possible to compute them, this is in most cases an effort without compensation. In fact the DM can not make a well fundamented choice if he is faced with a large set of solutions, most of them with just slightly different values of the objective functions. Besides, the best trade off solution may be a point belonging to a Pareto optimal surface which is not an extreme point. The consideration of this possibility makes even more difficult to calculate and present the results to the decision maker. These difficulties led us to study the possibility of applying interactive methods to this type of problem. Taking into account some general views expressed by B. Roy [3] a new method named "TRIMAP" was developed (for details, see [1]). The method TRIMAP is based upon the progressive and selective "learning" of the set of Pareto optimal solutions. Here the aim is not to "converge" on any "best compromise" solution, but to help the DM eliminating the subsets of the Pareto optimal solutions which are of no interest to him. The method combines three main procedures: weighting space decomposition, introduction of constraints on the objective functions space and/or on the weighting space. Furthermore, the introduction of constraints on the objective functions values can be translated into the weighting space.

In each step of the man/machine dialogue (when a new phase of calculations is being prepared) the analyst just asks the decision maker to give some indications about the directions to carry on the search for the Pareto optimal solutions. At this point, restrictions on the values of some of the objective functions may also be introduced by the decision maker. The process only ends when the D. M. considers to have "sufficient" knowledge about the set of Pareto optimal solutions. There are no irrevocable decisions along the process being always admissible to go "backwards". The method is not too demanding with respect to the information required from the decision maker in each iteraction, the main concern being to have a simple and flexible procedure. The TRIMAP is intended for tricriteria problems. This fact although being a limitation allows for the use of graphical means which are particularly convenient for the dialogue with the DM.

Tricriteria Linear Programming :

1. Problem Formulation

$$\underset{\underline{x} \in F}{\text{Max}} \ C\underline{x}$$

$$C = [\ \underline{c}_1\ ,\ \underline{c}_2\ ,\ \underline{c}_3\ ]'$$

$$F = \{\ \underline{x} \in \mathbb{R}^n\ :\ A\underline{x} = \underline{b}\ ,\ \underline{x} \geq 0\ \}$$

where "$\underline{c}_1$, $\underline{c}_2$, $\underline{c}_3$" are "n*1" column vectors, "A" is a "m*n" matrix and "$\underline{b}$" is a "m*1" column vector.

Without loss of generality we will admit that "A" has rank "m" and that all the constraints are converted into equalities. "Max" denotes the operation of finding the set of Pareto optimal solutions.

2. Pareto optimal solutions set definition :

Given a vectorial function

$$\underline{f}(\underline{x}) = C\underline{x}\ :\ \mathbb{R}^n \rightarrow \mathbb{R}^3$$

where $\underline{f}(\underline{x})$ is defined in F.

The set of Pareto optimal solutions is defined by

$$F_N = \{\ \underline{x} \in F\ :\ \underline{f}(\underline{x}_1) \geqq \underline{f}(\underline{x}) \Rightarrow \underline{x}_1 \notin F\ \}$$

where:

$$\underline{f}(\underline{x}_1) \geqq f(\underline{x}) \text{ iff } \begin{cases} \underline{f}(\underline{x}_1) \geq \underline{f}(\underline{x}) \\ \underline{f}(\underline{x}_1) \neq \underline{f}(\underline{x}) \end{cases}$$

$$\underline{f}(\underline{x}_1) \geq \underline{f}(\underline{x}) \text{ iff } f_i(\underline{x}_1) \geq f_i(\underline{x})\ ,\ i = 1,2,3$$

From the definition it follows that the satisfactory compromise solution must be searched in this set. For details see [4,7].

The Pareto optimal solutions will also be designated as non-dominated solutions.

3. TRIMAP Fundamentals

The interactive process begins with the automatic obtainment of a number of Pareto optimal extreme points. Next, a calculus phase is prepared through a dialogue phase with the decision maker. These two phases are repeated until the decision maker considers to know the sufficient about the Pareto optimal solution set.

Beyond the calculus phase and the dialogue phase between the analyst/computer and the DM we want to stress the importance of a "sensitizing phase" on the nature, potencialities and limitations of the method.

The calculus process of Pareto optimal feasible polyhedron extreme points used in the TRIMAP consists in solving problems of the type:

$$\max_{\underline{x} \in F} \{ w_1 f_1(\underline{x}) + w_2 f_2(\underline{x}) + w_3 f_3(\underline{x}) \} \qquad (1)$$

where

$$\begin{cases} \Sigma w_i = 1 \\ w_i > 0 \end{cases}, \qquad i = 1, 2, 3$$

Note that $\underline{x} \in F$ is a Pareto optimal solution if and only if it is an optimal solution of (1).

Furthermore, it must be remarked that only two weights are independent variables. For instance,

$$w_3 = 1 - w_2 - w_1 .$$

So, the graphical representation of the set of

$$\underline{w} = [\ w_1\ ,\ w_2\ ,\ w_3\ ]$$

that give origin to each Pareto optimal extreme point can be achieved through the triangle decomposition. In this paper we always consider the triangle of fig.1.

In general, the triangle will be completely filled when all the Pareto optimal feasible polyhedron vertices have been found.

The main purpose will be to do progressive and selective triangle filling. In each step the DM will be called to decide whether to study or not the solutions corresponding to not yet searched triangle regions. We want to prevent the exaustive search of regions with close objective function values (very often found in real case studies). It is also possible to eliminate triangle regions by imposing limits on the values of the objective functions (see the example in part II).

The introduction of additional constraints can also be used to get some Pareto optimal solutions that are not vertices of the feasible polyhedron (see [1]).

The selection of the weights may be automatic (in this way it is possible to get a certain number of Pareto optimal extreme points, the weights that originated them being well distributed on the triangle) or manual (the weights selection is made by the DM's indication of the triangle zones not yet filled, for which he thinks to be important the continuation of the search). Furthermore, if the functions have not in the same units, it is convenient to do its normalization.

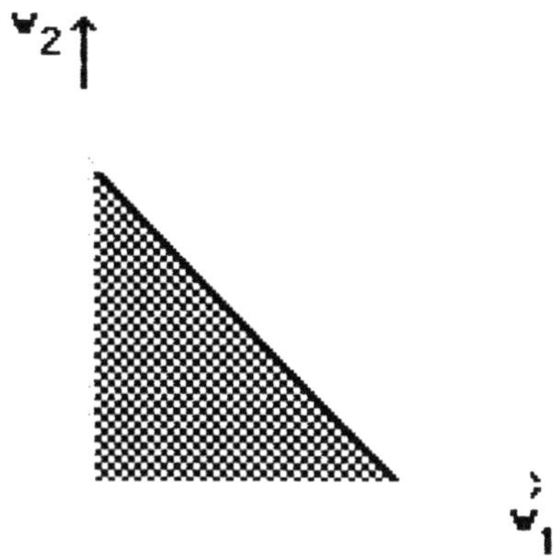

Fig. 1

4. Results presentation

In each interaction two graphs will be presented to the decision maker. The first one is the triangle filled with the regions corresponding to each of the already known Pareto optimal vertices. Eventual constraints on the variation of "$\underline{w}$" will also be presented. The second one presents a projection on the plan "f1 f2" of the Pareto optimal solution already computed. The area of each region of the triangle, corresponding to each Pareto optimal extreme point already computed, is presented. The Tchebycheff distance (L_∞) from each Pareto optimal extreme point to the "ideal solution" can also be presented. The so called "ideal solution" is one that would optimize all the objective functions. Of course it is not feasible in general !

5. Software implementation

The first TRIMAP version is running on a Apple Macintosh

Plus microcomputer. The whole program is written in Pascal language.

The friendly Macintosh user interface is specially suitable for taking advantage of the TRIMAP interactive capabilities. Macintosh facilities (including the mouse, windows, pull-down menus, etc.) are incorporated into the program making easier the DM tasks.

The Macintosh philosophy "What You See Is What You Get" allows users with little experience with computers to use TRIMAP, through the mouse selecting of actions (setting a lower bound for an objective function, getting all the information about a given solution, pointing into a not yet filled triangle region to choose a new weights set, etc.).

The program is intended both for real world studies and educational purposes. The Macintosh implementation is particularly suitable for educational purposes. In what is concerned with the study of real world problems we are applying the method to a power systems expansion planning problem (a first version of this work is in [2]).

At present we are working in the implementation of TRIMAP in a more powerful computer since in medium size problems the Macintosh Plus version is too slow, being impossible to apply this version to large dimension problems.

PART II

TRIMAP as a teaching tool

Introduction

We'll use an example for showing the main potentialities of the TRIMAP package, giving particular attention to the details thought to be more relevant to the teaching of multicriteria linear programming. Note that we'll be dealing with a non-degenerate example. Otherwise special cautions should be considered. It should be emphasized that the TRIMAP version implemented in Macintosh Plus is particularly suitable for educational purposes. This has been experienced in the discipline "Decision Support Methods" at the Department of Electrical Engineering Science of the University of Coimbra (Portugal). In the original paper on TRIMAP [1] it was not drawn the attention to some details which are relevant for the teaching purposes. We could refer, for example, the fact that when introducing constrains in the admissible region, some non-dominated solutions in the new problem may exist which are dominated solutions with respect to the original admissible region.

Teaching MCDA Using TRIMAP -- A Guided Tour

Before doing a guided tour of TRIMAP (through an appropriate example) we would like to draw the attention to the following points :

1. Although out of the scope of this paper, we should refer that the TRIMAP package can be improved for educational objectives if some changes and the introduction of some new procedures are carried out. This will enable

the teaching of several interactive methods (as an example we could refer the methods of Zionts-Wallenius and the "Steuer's Contracting Cone Method") by making the most of the graphical interface capability.

2. The utilisation of the mouse and of the various available "menus" which will be mentioned in the example's presentation (see the next paragraph) will enable the teacher to stimulate the students to carry out several experiments with simple problems, as for example by studying the variations in the decomposition of the weighting space resulting from variations in the relative dimensions of the objective functions gradients, in the geometry of the admissible region or in the criteria cone. Since it is not possible to be exaustive in this domain we'll refer as an example some conclusions which may be obtained from the total decomposition of the weights space (for a very simple example presented in [7] - fig. 2a) and from the decomposition of the weights space (for an example that only differs from the previous one by the function f3(x)).

$$\underset{\underline{x} \in F}{\text{Max}} \begin{cases} f_1 = -x_1 + 2x_2 \\ f_2 = 2x_1 - x_2 \\ f_3 = 2x_1 + x_2 \end{cases} \qquad \underset{\underline{x} \in F}{\text{Max}} \begin{cases} f_1 = -x_1 + 2x_2 \\ f_2 = 2x_1 - x_2 \\ f_3 = 3x_1 - x_2 \end{cases}$$

$$F = \{ \underline{x} \in \mathbb{R}^2 : x_1 \le 4 , x_2 \le 4 , x_1 + x_2 \le 7 , -x_1 + x_2 \le 3 , x_1 - x_2 \le 3 , x_1 \ge 0 , x_2 \ge 0 \}$$

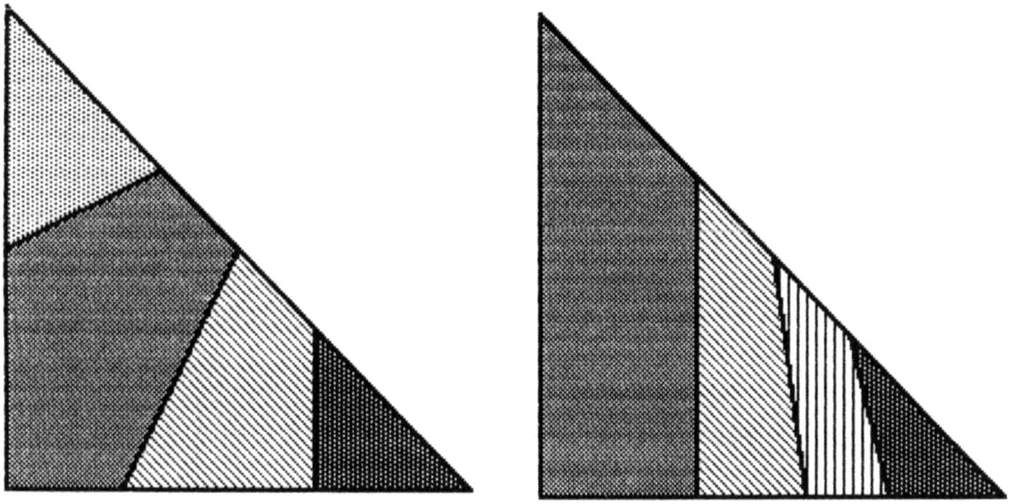

(a) (b)

Fig. 2

From fig. 2a it may be concluded that the criteria cone is plane and from fig. 2b it may be immediately said that the optimum of $f_2(\underline{x})$ and $f_3(\underline{x})$ is the same. From the first example the students may verify that there are cases where the results from the decomposition of the triangle enable to obtain non-dominated solutions for certain choices of the weights, which empirically did not seem reasonable. For example, if $w_1 = 0.4$, $w_2 = 0.6$, $w_3 = 0$, the obtained non-dominated solution corresponds to the optimum of $f_3(\underline{x})$. The students must be led to think of the reasons why these situations may happen. The TRIMAP software will enable them to carry out the experiments they find necessary in a rapid way.

Many other things may be "discovered" by the students by exploring this software. This may be achieved not only from the information contained in each graphic but also from the comparative study resulting from the exploration of the two graphics in the course of the interactive procedure.

Example

To illustrate the potentialities of TRIMAP, we will use the example (as Zionts in [7]):

$$\underset{\underline{x}\in F}{\text{Max}} = \underset{\underline{x}\in \mathbb{R}^4}{\text{Max}} \left\{ \begin{bmatrix} 3 & 1 & 2 & 1 \\ 1 & -1 & 2 & 4 \\ -1 & 5 & 1 & 2 \end{bmatrix} \begin{bmatrix} x_1 \\ x_2 \\ x_3 \\ x_4 \end{bmatrix} \right\}$$

such that

$$A\underline{x} \leq \underline{b}$$

where

$$A = \begin{bmatrix} 2 & 1 & 4 & 3 \\ 3 & 4 & 1 & 2 \end{bmatrix}$$

$$\underline{b} = \begin{bmatrix} 60 \\ 60 \end{bmatrix}$$

"x_5" and "x_6" are slack variables. "Max" denotes the operation of finding the Pareto optimal solutions set.

Guided Tour of TRIMAP

1. We begin by generating a certain number of Pareto optimal feasible polyhedron vertices using weights well distributed in the weighting space. The program generates automatically the Pareto optimal solutions corresponding to the optimum of each objective function. The graphics of fig. 3 are then obtained.

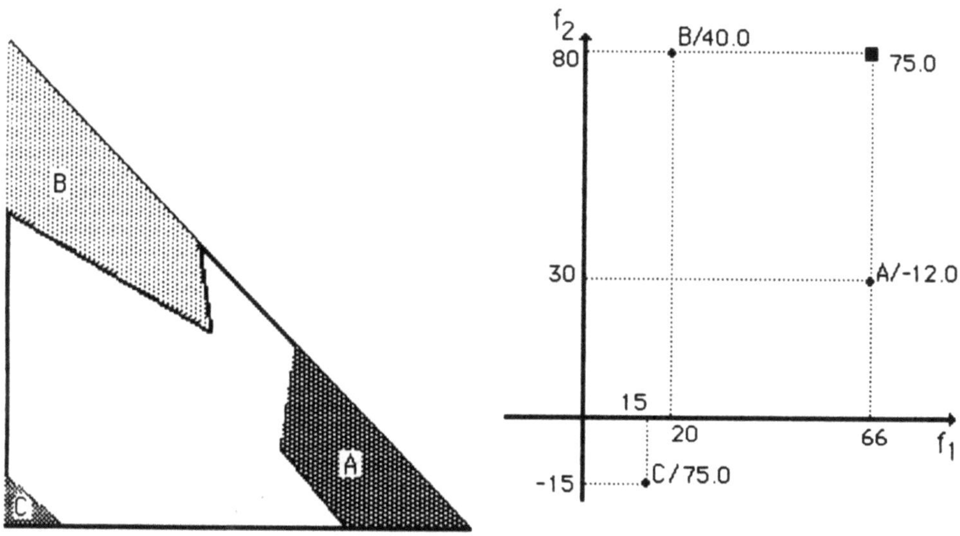

(a) (b)

Fig. 3

From the educational point of view, we will draw the attention to the following points:

(a) In a progressive and selective "learning" method, it is strategically important to begin the search for the Pareto optimal solutions by starting with very different solutions with respect to the objective functions values. This because it is important to obtain as quickly as possible the maximum information which will enable the Decision Maker (among other things) to establish eventual limitations for the objective functions values. We'll return later to this subject.

(b) From fig.3a it may be immediately concluded that there are no alternative optima of $f_1(\underline{x})$, $f_2(\underline{x})$, $f_3(\underline{x})$ which are Pareto optimal solutions.

(c) We can avail ourselves of the opportunity to draw the attention of the students to the underlying ideas associated with the methods for generating the Pareto optimal extreme points of the feasible polyhedron corresponding in general to the coverage of the whole triangle. When the specialised simplex method for multicritera problems was developed in the 70's, attention was drawn to the fact that the versions that were based in the decomposition of the parametric space presented several disadvantages, namely resulting from:

(i) When all the Pareto optimal feasible vertices have been found, the triangle filling may not be complete if there are degenerate solutions.

(ii) The triangle filling may be complete without all the Pareto optimal vertices being known, if there are several alternative solutions which optimize "$w_1f_1(\underline{x}) + w_2f_2(\underline{x}) + w_3f_3(\underline{x})$" for the same sub-area of the triangle.

As for TRIMAP we should emphasize at this point:

(1) Even in the above mentioned cases, the essential of the method can be used.

(2) From the pedagogical point of view, situations of this

type may be observed in real time. For example, if in a given problem there is an area of the weights space corresponding to two vertices, then, when the first one is calculated, that area is covered by a "texture" which changes when the calculations for the second one are complete.

We should also draw the student's attention to the utility function methods (an example of which will be a weighted sum of $f_1(x)$, $f_2(x)$ and $f_3(x)$ and to the interactive methods such as the one by Zionts and Wallenius (see [7]) where it is assumed from the outset, the existence of an implicit utility function of the DM, the interactive process being built with the aim of converging to the optimum of such a function.

2. Before continuing the guided tour, let us consider the information that the DM may obtain from each solution, if he uses for that purpose the item "indices" of the menu "special". For example, in fig. 4 solution "A" was selected. Beyond basic variables values and the objective functions values, we obtain the % of the global area of the triangle covered by that solution and the corresponding "L_∞" value.
3. Next we'll enter a dialogue phase with the DM. Since we are in an initial phase of the interactive procedure, the DM will in general make a choice for a new calculation stage. However the DM may choose one of the following options at any phase of the dialogue:

(a) To terminate the procedure, if he considers to have sufficient knowledge about the non-dominated optimal solutions set.

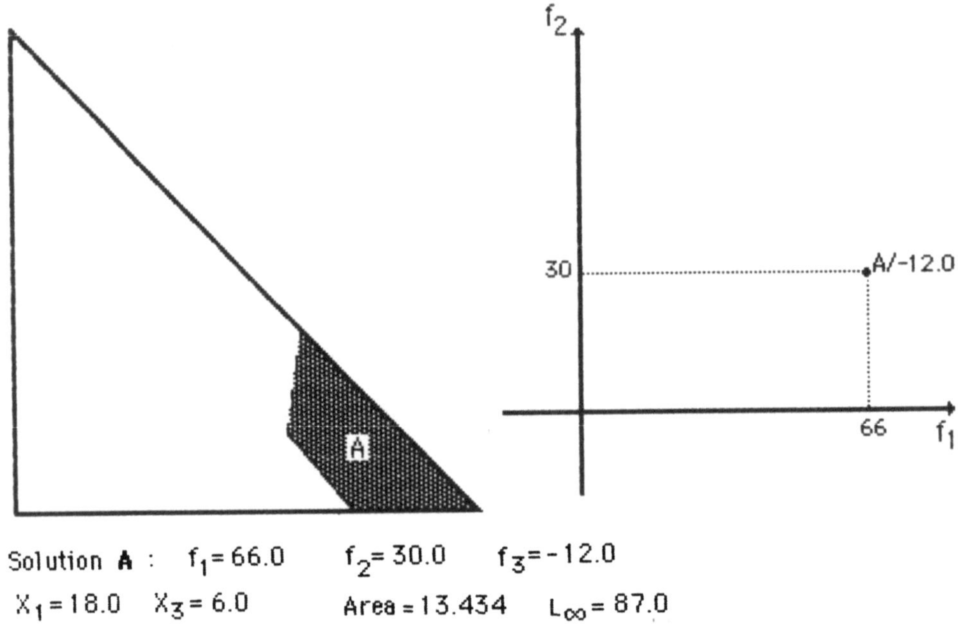

(a) (b)

Fig. 4

(b) To introduce new constraints, either on the weights space, or in the objective functions values, thus only considering in subsequent calculations the values of "$\underline{w}$" satisfying those constraints.

(c) To proceed directly with the calculation of a new Pareto optimal extreme point.

4. Let us now suppose that the above mentioned option (c) has been chosen. There are then several possibilities:

(a) To contruct a new weighted objective function for which a constant cost plane passes through the previous calculated points A, B, C. This procedure (item composition of the "menu" special) may be used for any set of three P.O.E.P.

(Pareto optimal extreme points) previously computed when the obtained weights are strictly positive. Otherwise this is ensured by an automatic change of the weights with the least perturbation of the weighted objective function gradient.

(b) To use directly the item "by mouse" from the menu "weights" in order to select a not yet occupied point of the triangle.

(c) To select a new system of weights by using the item "by numerical values" from the menu "weights".

Note that the utilisation of option (a) by students makes them realize that any objective function generated in that way, leads to one (or more) Pareto optimal solution(s).

In the guided tour we are presenting, the option has been

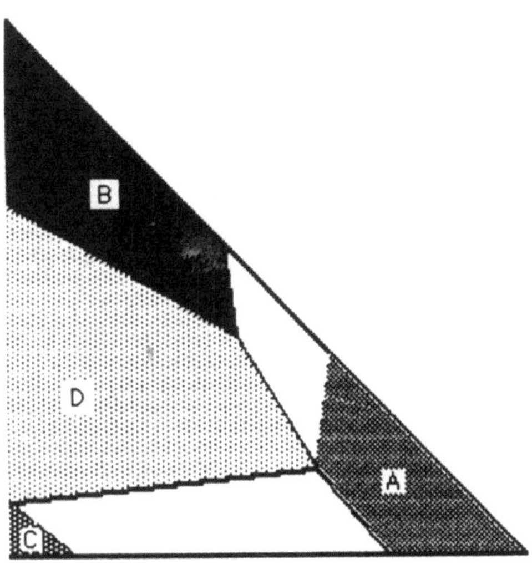

(a)

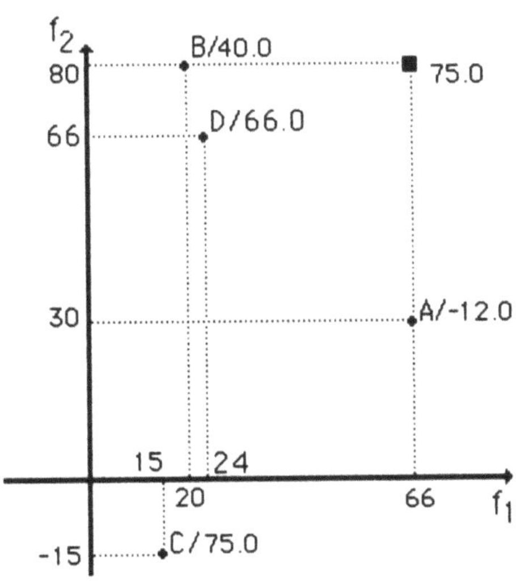

(b)

Fig. 5

made for continuing the triangle filling. This was done by choosing $w_1 = w_2 = w_3 = 1/3$ and leads to the solution "D" in fig. 5.

A dialogue phase is then started....

5. We'll now draw the attention to one of the most important potentialities of TRIMAP. As referred in the introduction, there are many real problems where the number of Pareto optimal extreme points, with very close values for the objective functions, is relatively high. This type of situation must be dealt with in a specific way in order to prevent the calculation of an excessive number of Pareto optimal extreme points. The comparative observation of the weights space and the objective functions space as allowed by TRIMAP, gives us the possibility of detecting those situations thence preventing the search for the solutions corresponding to the triangle areas associated with that type of conditions.

As an example refer what occurs in a simplified power systems expansion planning problem which has been used for testing TRIMAP. This problem was modeled as a tricriteria linear programming problem with 28 constraints and 99 decision variables. The use of TRIMAP led to the triangle decomposition in fig. 6. The hypothetical DM considered that the search for 16 solutions gave him a satisfactory knowledge of the Pareto optimal extreme points set. It should be noticed that an exaustive search leads to about 100 Pareto optimal extreme points (see fig. 6).

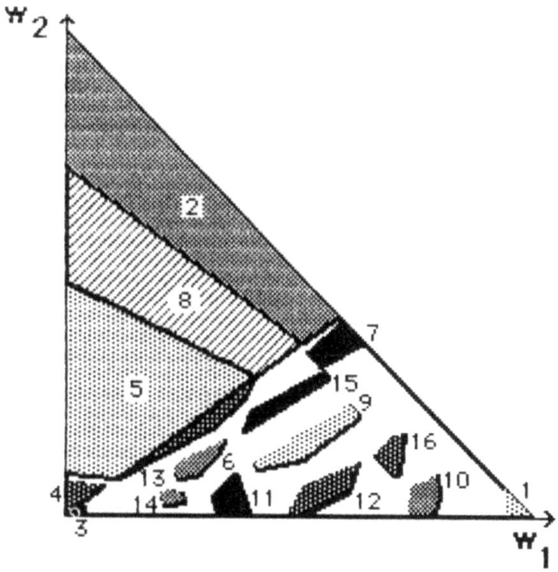

Fig. 6

From the educational point of view, we should emphasize that it was observed that the students practice with TRIMAP leads them to progressively "discover" -- by using the two graphical representations at their disposal -- the form of the set of non-dominated solutions in the 3-dimensional space "f_1 , f_2 , f_3" .

6. Let us return to the "guided tour". If for this particular example we continue the triangle filling procedure, the triangle will end completelly filled, as in fig. 7. In this case all the Pareto optimal extreme points were found. It must be emphasized that in general this is not a TRIMAP goal.

With respect to the teaching side , we may say that the comparison of the two graphics of fig. 7 is particularly interesting:

a. Note that the projection of the surfaces signalized in fig. 7b corresponds to points in fig. 7a.

b. By simple inspection of fig. 7a, we may conclude that the surfaces "B D F" and "A E D F" are non-dominated, while surface "C D E" is dominated, only the edges "DE" and "CE" being non-dominated.

c. On the other hand, the non-dominated vertices represented in fig. 7b correspond to sub-areas of the triangle of fig. 7a.

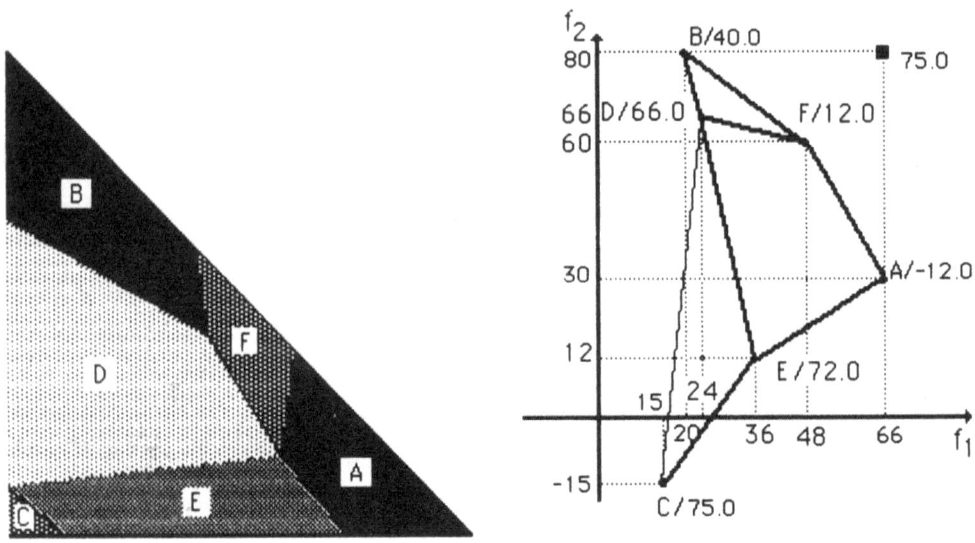

(a) (b)

Fig. 7

The student must be stimulated in thinking of justifying these facts.

We should also note that from simple observation of fig. 7, it may be seen why the variable x_5 is not efficient in the sense referred by the method of Zionts & Wallenius, despite being possible to go from the non-dominated extreme point "D" to "C" by introducing the variable x_5 into the basis (see [5]). The reason for this is the fact that the edge "DC" is dominated at the exception of the points "D" and "C".

7. As previous mentioned, whenever a dialogue phase occurs, the question is risen of eventually introducing bounds for the objective functions values. In this paragraph we'll illustrate the procedure which takes place when the DM decides to introduce such bounds (in the present case there will be lower bounds, since the objective functions are to be maximized) in one or more objective functions by using the item "objective functions space" of the menu "constraints".

Let us show what happens when the DM establishes a lower bound for an objective function value.

Let it be "$f_2(x) \geq 28$". The following auxiliary problem is constructed:

$$\max_{\underline{x} \in F_a} \underline{f}(\underline{x})$$

where

$$F_a = \{ \underline{x} \in F : f_2(\underline{x}) \leq 28 \}$$

We now select the vertices of F_a (satisfying the condition that $f_2(\underline{x}) \leq 28$ is active) which are Pareto optimal extreme points and the corresponding areas in the triangle are computed ("X", "Y" and "Z"). For this purpose we just calculate the alternative optimal solutions of $f_2(\underline{x})$

and select the non-dominated ones -- see fig. 8.

Note that the point, which corresponds to the intersection of the edge "DC" with $f_2(\underline{x}) = 28$ -- see fig. 8, is dominated by "Z". Going backwards to the original problem, the consideration of the constraint $f_2(\underline{x}) \geq 28$ corresponds to limitate the weights systems search to the shadowed area in fig. 9.

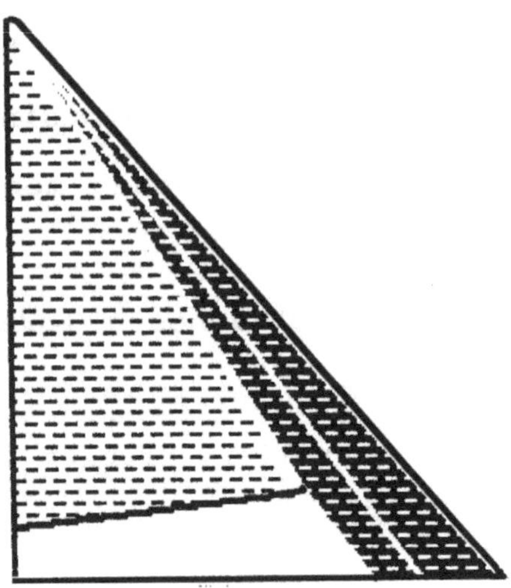

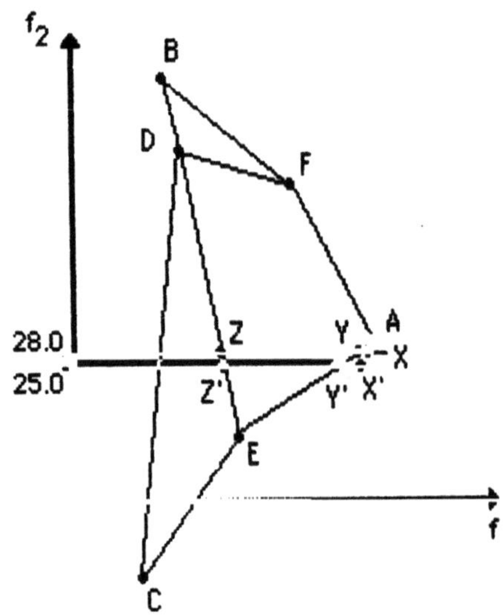

Fig. 8

	f_1	f_2	f_3
X	64.8	28	-13.6
Y	62.7	28	- 2.7
Z	32.4	28	70.2

	f_1	f_2	f_3
X'	63	25	-16
Y'	57.7	25	11.3
Z'	31.1	25	70.6

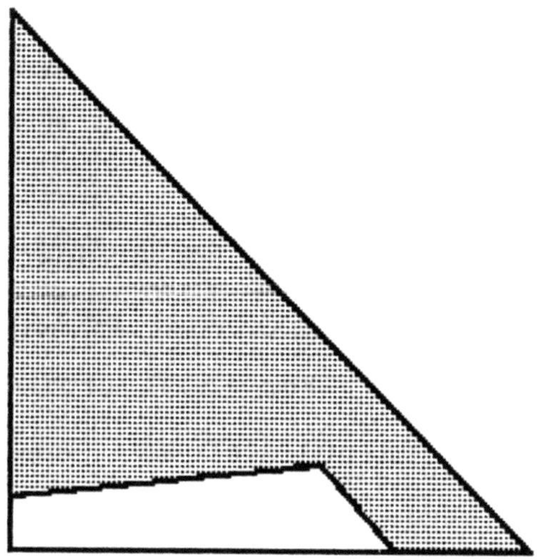

Fig. 9

This facility is rather important, since it seems to us that asking the DM to establish those lower bounds is something that he easilly understands. This type of option can be made at any stage of the decision procedure by using the available information which is easy to assimilate by the DM.

On the other hand, we should note that the menu "special" has an item, called "backwards", which enables to go backwards at any time.

We must also refer that the value "L_∞" which is calculated for each solution, as previously mentioned, may be a very useful indicative parameter for establishing the bounds in question.

If we search the whole area marked in fig. 9 we obtain fig. 10.

Before finalizing this paragraph, note that fig. 8 is quite convenient for showing the students the aspect of the triangle decomposition near a vertice when there are alternative optima for one of the objective functions which are Pareto optimal.

8. The DM may also get to know some non-dominated solutions of the original problem with $f_2(\underline{x}) = 28$ by using the item "cuts" of the menu "special". For this purpose the Pareto optimal points belonging to non-dominated edges for which $f_2(\underline{x}) = 28$ are calculated. It should be noted here that there may exist -- and that is the case of the present example -- non-dominated vertices for the auxiliary problem, where $f_2(\underline{x}) = 28$, which correspond to dominated solutions for the original problem. In fig. 8 this ambiguity is evidenciated by making a new cut very close to the first one and then checking which of the pairs of points "X,X′","Y,Y′"

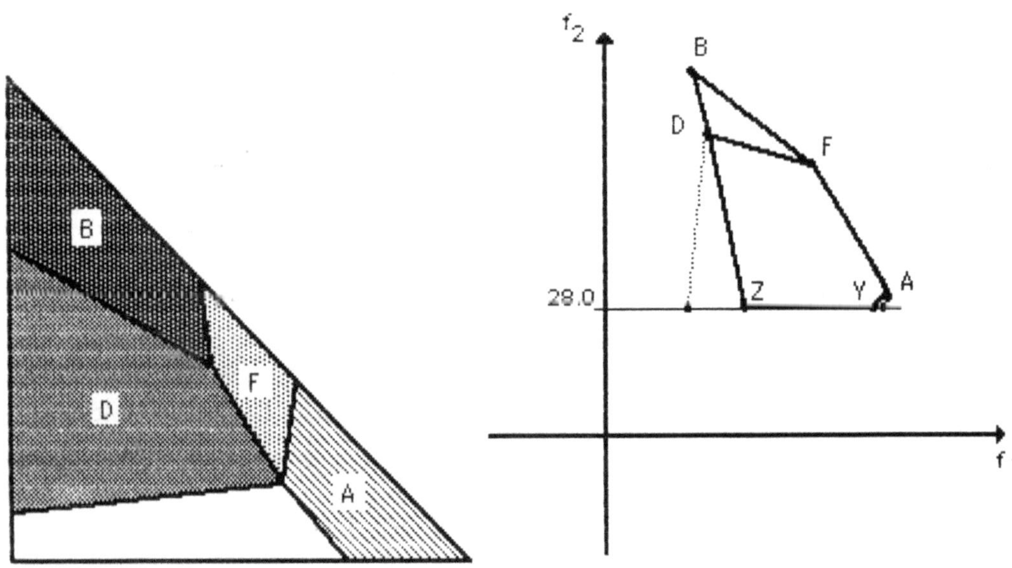

(a) (b)

Fig. 10

and "Z,Z'" are non-dominated among themselves. In this case, it can be verified from the values shown in fig. 8 that X dominates X'. Note that the triangle decomposition of fig. 8 implies that $f_2(\underline{x}) = 28$ crosses two nondominated edges of the original problem. It may then be concluded that only "Y" and "Z" are non-dominated in relation to the original problem.

If for example we wish to know some non-dominated solutions in the surface to which "Y" and "Z" belong and maintain $f_2(\underline{x}) = 28$, we'll just have inspect in fig. 8 which values may be fixed for $f_1(\underline{x})$ (in this case these values will be in the interval from 32.4 to 62.7). Once the desired value for $f_1(\underline{x})$ (for example $f_1(\underline{x}) = 40$) has been fixed, we'll use the item "cuts" of the menu "special" for fixing $f_1(\underline{x}) = 40$ and $f_2(\underline{x}) = 28$ and then optimize a weighted sum of the functions $f_1(\underline{x})$, $f_2(\underline{x})$ and $f_3(\underline{x})$. This is done in such way that the surface we are inspecting is a constant cost plane for this function and the domain is the original one with the additional constraints $f_2(\underline{x}) = 28$ and $f_2(\underline{x}) = 40$.

Under the practical point of view, it is of major interest the possibility of calculating the Pareto optimal solutions which are not vertices of the admissible polyhedron. For example, when planning new electrical power plants, as in the case study referred in [2], there are non-dominated surfaces which correspond to sharp variations in the objective functions values. Since this type of situation is common in real problems, it is important to enable searching for solutions in those surfaces, distinct from the vertices, if requested by th DM. Otherwise the search for satisfactory compromise solutions could be significantly impaired.

Obviously the subjects treated under paragraph 7 and 8 are very rich in terms of themes for discussion and elucidation of the students.

Final remarks

In the following remarks we emphasize once more not only the possible advantages of a new philosophy on interactive methods developed in [3] and explored by us in TRIMAP package, but also the potentialities of this package for teaching purposes.

a) The TRIMAP is based on a progressive and selective "learning" of the Pareto optimal solutions set. It is not intended to "converge" to any "best" compromise solution, but to help the DM to eliminating the Pareto optimal solutions sub-set in which he is not interested at all. There are no irrevocable decisions during the whole process and the DM is always allowed to go "backwards" at a later interaction.

So, in each interaction (when preparing a new calculus phase) the analyst limits himself to ask the DM some indications on the direction where to continue the search for Pareto optimal solutions, as well as eventual introduction of constraints. The process only finishes when the DM considers "to know sufficient" about the Pareto optimal solutions set.

Using B. Roy's terminology, we may say that in TRIMAP the "convergence is replaced by the "creation" and that the interactivity process is a "building process" and not the discovery of some pre-existent utility function.

b) A problem always present in the interactive methods is the limited capacity of the human being processing the information. The TRIMAP is dedicated to problems with three objective functions. Although being a limitation, this allows the use of graphical means specially suitable for the "dialogue" with the decision maker. This undoubtfully enlarges the decision maker's capacity to process the information, by simplifying the dialogue phase.

The good-working of TRIMAP method implies, beyond the traditional phases of calculus and dialogue with the DM, a phase of bringing him to sensibility with the method potentialities and limitations, as well as with the consequences of the decisions to take.

c) As for the application of the TRIMAP package for teaching purposes, I think the text has made clear that it can be very useful, not only because the nature of the TRIMAP favours the understanding of many of the questions raised by multicriteria linear programming problems, but also because the type of information available to the DM

(namely through the graphical means) allows for the teaching of the basics of multicriteria programming in a very attractive way. Finally it must be referred that we are preparing a Teaching Manual to be presented with an educational TRIMAP diskette that will be made available to the scientific community of the area, during the international meetings programmed for the summer 88. In this package some other extra facilities will be introduced as for example the possibility of exploring the information contained in the reduced costs matrix, relative to the objective functions $f_1(\underline{x})$, $f_2(\underline{x})$ and $f_3(\underline{x})$ of the simplex tableau, corresponding to each of the calculated non-dominated extreme points.

References

[1] Clímaco, J. C. N. and Henggeler Antunes, C. "TRIMAP -- An Interactive Tricriteria Linear Programming Package", to appear in Foundations of Control Engineering, vol. 12, nº 3, 1987

[2] Clímaco, J.; Antunes, C. H.; Martins, A.; Marta, J. and Almeida, A. T. "Power Generation Expansion Planning Using a Tricriteria Interactive Approach" Proceedings of the MELECON - IEEE, Italy 1987

[3] Roy, B. "Meaning and Validity of Interactive Procedures as Tools for Decision Making" Cahiers LAMSADE nº 62, 1985

[4] Steuer, R. "Multiple Criteria Optimization : Theory, Computation, and Application" Wiley, 1986

[5] Samblancks, S.; Depraetere, P. and Muller, H.
"Critical Considerations Concerning the Multicriteria Analysis by the method of Zionts and Wallenius"
EJOR, vol. 10, nº 1, 1982

[6] Clímaco, J. C. N.; and Traça - Almeida, A.
"Multiobjective Power Systems Generation Planning"
Proc. of the Third International Conference on Energy Use Management
Pergaman Press, 1981

[7] Zionts, S.
"Multicriteria Mathematical Programming : An Updated Overview and Several Approaches"
In this publication

THE TCHEBYCHEFF PROCEDURE OF INTERACTIVE MULTIPLE OBJECTIVE PROGRAMMING

Ralph E. Steuer
Department of Management Science & Information Technology
Brooks Hall
University of Georgia
Athens, Georgia
USA 30602

Abstract

Reflecting its most recent refinements, this paper provides a detailed description of the interactive Tchebycheff procedure of multiple objective programming. The paper discusses eight prerequiste topics that are fundamental to the understanding of the Tchebycheff method, followed by an algorithmic specification of the procedure. Also, the paper shows how several additional features can be optionally incorporated into the algorithm.

1. INTRODUCTION

Consider the <u>multiple objective program</u>

$$\begin{aligned} &\max \; \{f_1(x) = z_1\} \\ &\quad\vdots \\ &\max \; \{f_k(x) = z_k\} \\ &\text{s.t.} \quad x \in S \end{aligned}$$

in which S is the feasible region in <u>decision space</u>. Let $Z \subset \mathbb{R}^k$ be the feasible region in <u>criterion space</u> where $z \in Z$ if and only if there exists an $x \in S$ such that $z = (f_1(x), \ldots, f_k(x))$. <u>Criterion vector</u> $\bar{z} \in Z$ is <u>nondominated</u> if and only if there does not exist another $z \in Z$ such that $z_i \geq \bar{z}_i$ for all i and $z_i > \bar{z}_i$ for at least one i. The set of all nondominated criterion vectors is designated N and is called the <u>nondominated set</u>. A point $\bar{x} \in S$ is <u>efficient</u> if and only if its criterion vector $\bar{z} = (f_1(\bar{x}), \ldots, f_k(\bar{x}))$ is nondominated. The set of all efficient points is designated E and is called

NATO ASI Series, Vol. F56
Multiple Criteria Decision Making and Risk Analysis
Using Microcomputers. Edited by B. Karpak and S. Zionts

the efficient set.

In the world of interactive multiple objective programming, a number of procedures have been proposed. We have STEM by Benayoun, de Montgolfier, Tergny and Larichev (1971), the Zionts-Wallenius procedure (1976 and 1983), Wierzbicki's Reference Point procedure (1980 and 1986), the Tchebycheff method of Steuer and Choo (1983), VIG by Korhonen and Laakso (1986), and others. The concensus is that a spectrum of interactive procedures is necessary because the procedure to use on a given application is often problem dependent.

The purpose of this paper is to describe, in detail, only the Tchebycheff procedure. Although the Tchebycheff procedure is presented in Steuer and Choo (1983) and Steuer (1986), this paper contains all recent refinements, and provides, as of this writing, the most modern statement of the procedure available.

2. IDEA OF TCHEBYCHEFF PROCEDURE

The idea of the Tchebycheff procedure is to sample a sequence of progressively smaller subsets of the nondominated set N until we locate a final solution (a solution close enough to being optimal to terminate the decision process).

The Tchebycheff method is a weighting vector space reduction approach. This is because, by working with sequences of progressively smaller subsets of weighting vector space

$$\Lambda = \{\lambda \in \mathbb{R}^k \mid \lambda_i \in (0,1),\ \sum_{i=1}^{k} \lambda_i = 1\}$$

we are able to sample sequences of progressively smaller subsets of N. On the first iteration, a group of dispersed λ-vectors from all of weighting vector space

$$\Lambda^{(1)} = \{\lambda \in \mathbb{R}^k \mid \lambda_i \in (\ell_i^{(1)},\ \mu_i^{(1)}),\ \sum_{i=1}^{k} \lambda_i = 1\}$$

$$\text{where } (\ell_i^{(1)},\ \mu_i^{(1)}) = (0,\ 1) \text{ for all } i$$

is used to sample the (entire) nondominated set N. On iteration h, where $h \geq 2$, a group of dispersed λ-vectors from

$$\Lambda^{(h)} = \{\lambda \in \mathbb{R}^k | \ \lambda_i \in (\ell_i^{(h)}, \mu_i^{(h)}), \ \sum_{i=1}^{k} \lambda_i = 1\}$$
$$\text{where } (\ell_i^{(h)}, \mu^{(h)}) \subset (0,1) \text{ for all } i$$

is used to sample a subset of N. Convergence through a sequence of progressively smaller subsets of N to a final solution is controlled by iteratively tightening the $(\ell_i^{(h)}, \mu_i^{(h)})$ bounds so that $\Lambda^{(h)}$ is a smaller percentage of Λ than $\Lambda^{(h-1)}$.

Most interactive procedures are restricted by simplifying assumptions about the decision maker's (unknown) utility function, the objective functions, or the feasible region. In contrast, the Tchebycheff method has been designed for <u>any</u> type of multiple objective program. This means that the Tchebycheff method is not only applicable to multiple objective linear programs, but to integer and nonlinear multiple objective programs as well. Moreover, the theory of the Tchebycheff method is sufficiently robust for the method to operate an intelligent fashion on problems with nonconcave utility functions (as long as "more is always better than less of each criterion"). In this way, the Tchebycheff method provides an approach useful on problems often excluded by the other procedures.

3. PREREQUISITE TOPICS

Before presenting the algorithmic specification of the Tchebycheff method in Section 4, eight short topics are discussed. They are (a) decision space vs. criterion space, (b) unsupported nondominated criterion vectors, (c) payoff tables, (d) utopian criterion vectors, (e) normalization of the objectives, (f) the lexicographic sampling program, (g) dispersed groups of λ-vectors, and (h) objective function contours of the lexicographic sampling program.

(a) <u>Decision Space vs. Criterion Space</u>. Rather than in decision space, it is most helpful to study the Tchebycheff method in criterion space. To illustrate the relationship between a multiple objective program's representation in decision space and its representation in criterion space, consider

$$\max \{x_1 - 1/2x_2 = z_1\}$$
$$\max \{\qquad x_2 = z_2\}$$
$$\text{s.t.} \qquad x \in S$$

where S in decision space is given in Figure 1, and Z in criterion space is given in Figure 2. For instance z^4, which is the image of x^4, is obtained by plugging (3, 4) into the objective functions to generate $z^4 = (1, 4)$. In this way, Z is the image of S under the f_i. In Figure 1 we see that the efficient set E is the set of boundary points x^3 to x^5 to x^6, and the image of E in Figure 2 is N which consists of the set of boundary points z^3 to z^5 to z^6. Note that Z is not necessarily confined to the nonnegative orthant.

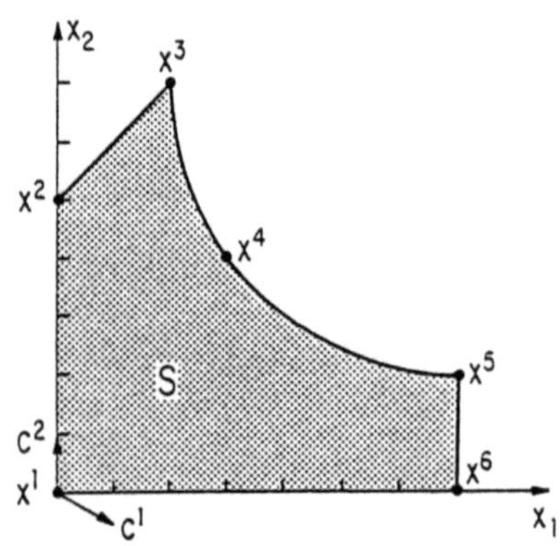

Figure 1. Representation in Decision Space.

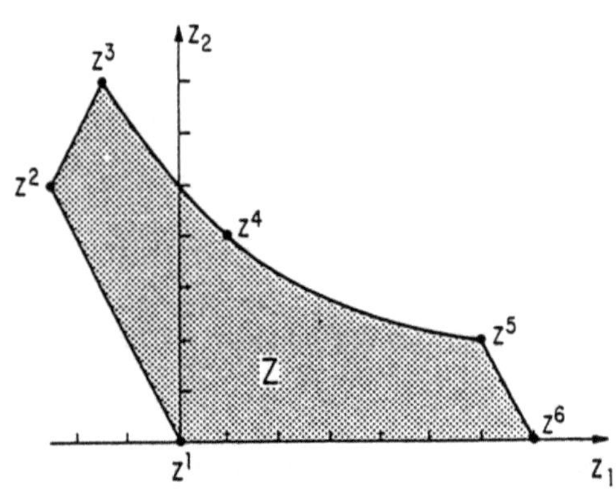

Figure 2. Representation in Criterion Space.

(b) Unsupported Nondominated Criterion Vectors. A nondominated criterion vector $\bar{z} \in Z$ is unsupported if and only if it does not maximize the weighted-sum program

$$\max \{\lambda^T z \mid z_i = f_i(x), \ x \in S\}$$

for any $\lambda \in \Lambda = \{\lambda \in \mathbb{R}^k \mid \lambda_i \in (0, 1), \ \sum_{i=1}^{k} \lambda_i = 1\}$. Unsupported nondominated criterion vectors are in "nonconvex depressions" of the nondominated set. In Figure 2, the set of unsupported nondominated criterion vectors is the set of criterion vectors from z^3 through z^4 to x^5, exclusive of z^3 and z^5. The set of supported nondominated criterion vectors is the set that consists of z^3 plus the line segment z^5 to z^6, inclusive.

(c) Payoff Tables. A payoff table is of the form

	z_1	z_2		z_k
z^1	z_1^*	z_{12}		z_{1k}
z^2	z_{21}	z_2^*		z_{2k}
			. . .	
z^k	z_{k1}	z_{k2}		z_k^*

where the rows are the criterion vectors resulting from individually maximizing the objectives. If one is worried about alternative optima, a way to assure that all row criterion vectors are nondominated is to lexicographically maximize the objectives.

The z_i^* entries along the main diagonal are the maximal criterion values for the different objectives over the nondominated set. The minimum criterion value in the i-th column of the payoff table is an estimate of the minimum criterion value of the i-th objective over N. Often these column minimums are used in place of the minimum criterion values over N because the minimum criterion values are difficult to obtain (see Weistroffer (1985), Dessouky,

Ghiassi and Davis (1986), and Isermann and Steuer (1988)).

d) Utopian Criterion Vectors. We will define a utopian criterion vector $z^{**} \in \mathbb{R}^k$ to be an (infeasible) criterion vector that strictly dominates every nondominated criterion vector. The components of z^{**} are formed as follows:

$$z_i^{**} = z_i^* + \epsilon_i$$

where the z_i^* are the maximal criterion values obtained from the main diagonal of the payoff table, and the ϵ_i are moderately small, computationally significant, positive values. An ϵ_i value that raises z_i^{**} to the smallest integer greater than z_i^* is normally sufficient.

(e) Normalization of the Objectives. One way to normalize the objectives is to equalize the R_i, where R_i is the difference between z_i^* and the minimum criterion value of the i-th objective over N. A way to roughly accomplish this purpose is to multiply each objective by its respective π_i factor

$$\pi_i = \frac{1}{\bar{R}_i} \left[\sum_{j=1}^{k} \frac{1}{\bar{R}_j} \right]^{-1} \qquad (3.1)$$

where $\bar{R}_i$ is the difference between z_i^* and the minimum value in the i-th column of the payoff table. Ideally, we would want to use R_i's instead of $\bar{R}_i$'s, but the R_i values are typically not available because of the difficulty in obtaining the minimum criterion values over the nondominated set.

(f) Lexicographic Sampling Program. For the lexicographic sampling program

$$\text{lex min } \{\alpha, \sum_{i=1}^{k} (z_i^{**} - z_i)\}$$

$$\text{s.t.} \quad \alpha \geq \lambda_i(z_i^{**} - z_i) \qquad i \in K$$

$$f_i(x) = z_i \qquad i \in K$$

$$x \in S$$

$$0 \leq \alpha \in \mathbb{R}$$

where z^{**} is a utopian criterion vector; $\lambda \in \Lambda = \{\lambda \in \mathbb{R}^k | \ \lambda_i \in (0,1),$

$\sum_{i=1}^{k} \lambda_i = 1\}$; and $K = \{1,2,\ldots,k\}$, we have the following two results:

(1) If $\bar{z} \in Z$ is a minimizing criterion vector solution of the lexicographic sampling program, $\bar{z}$ is nondominated.

(2) If $\bar{z} \in Z$ is nondominated, there exists a $\lambda \in \Lambda$ such that $\bar{z}$ uniquely solves the lexicographic sampling program.

Result (1) states that the lexicographic sampling program will only generate nondominated criterion vectors. Result (2) means that no nondominated criterion vector can hide from the lexicographic sampling program.

(g) Dispersed Groups of λ-Vectors. On each iteration h, the Tchebycheff method needs a group of dispersed λ-vectors from

$$\Lambda^{(h)} = \{\lambda \in \mathbb{R}^k \mid \lambda_i \in (\ell_i^{(h)}, \mu_i^{(h)}), \sum_{i=1}^{k} \lambda_i = 1\}$$

$$\text{where } (\ell_i^{(h)}, \mu_i^{(h)}) \subset (0, 1) \text{ for all } i$$

By dispersed we mean λ-vectors that are as evenly distributed over $\Lambda^{(h)}$ as far as it is practicable to compute. Such groups of dispersed λ-vectors can be obtained by using the LAMBDA and FILTER codes from Steuer (1983b). The LAMBDA code randomly generates λ-vectors from $\Lambda^{(h)}$. The FILTER code computes groups of dispersed representatives from the set of vectors inputted to it. Suppose we need 10 dispersed λ-vectors from $\Lambda^{(h)}$. Then the LAMBDA program would be asked to produce an overabundance of, say 200, randomly generated λ-vectors from $\Lambda^{(h)}$. Then FILTER would be called upon to select 10 λ-vectors that are widely dispersed among the 200. The most modern statement about the functions performed by LAMBDA and FILTER is found in Steuer (1986, Chapter 11).

(h) Objective Function Contours of the Lexicographic Sampling Program. The objective function contours of the first optimization stage of the lexicographic sampling program are L-shaped sets whose vertices lie along the line emanating from z^{**} in the direction

$$-\left[\frac{1}{\lambda_1}, \frac{1}{\lambda_2}, \ldots, \frac{1}{\lambda_k}\right]$$

In terms of these L-shaped contours, the first optimization stage determines the "L-shaped" contour closest to z^{**} that intersects Z. If there is only one feasible criterion vector on this L-shaped contour, this is the nondominated criterion vector returned by the lexicographic sampling program.

If there is more than one feasible criterion vector on the L-shaped contour, some may be dominated. In this case, the second optimization stage is employed to determine a feasible criterion vector on the L-shaped contour that is closest to z^{**} according to the L_1 metric. In this way, the lexicographic sampling program terminates with a criterion vector that is assured to be nondominated.

4. ALGORITHMIC SPECIFICATION

The interactive Tchebycheff procedure is now specified:

Step 1: Specify the number of solutions P to be presented to the decision maker per iteration, the number of iterations t the procedure is to run, and the Λ-reduction factor r.

Step 2: Let h = 0. Let $(\ell_i^{(1)}, \mu_i^{(1)}) = (0, 1)$ for all i.

Step 3: Construct a z^{**} utopian criterion vector.

Step 4: Let h = h + 1. Form

$$\Lambda^{(h)} = \{\lambda \in \mathbb{R}^k | \ \lambda_i \in (\ell_i^{(h)}, \mu_i^{(h)}), \ \sum_{i=1}^{k} \lambda_i = 1\}$$

where $(\ell_i^{(h)}, \mu_i^{(h)}) \subset (0, 1)$ for all i

Step 5: Obtain a group of 2P dispersed λ-vectors from $\Lambda^{(h)}$.

Step 6: Once for each of the 2P dispersed λ-vectors, solve the lexicographic sampling program

$$\text{lex min } \{\alpha, \sum_{i=1}^{k} (z_i^{**} - z_i)\}$$

$$\begin{aligned} \text{s.t.} \quad & \alpha \geq \lambda_i(z_i^{**} - z_i) & & i \in K \\ & f_i(x) = z_i & & i \in K \\ & x \in S & & \\ & 0 \leq \alpha \in \mathbb{R} & & \end{aligned}$$

<u>Step 7</u>: Present the P most different of the resulting criterion vectors to the decision maker for the selection of his or her most preferred, designating it $z^{(h)}$.

In Step 6, even though we only plan to present P solutions to the decision maker, we <u>oversample</u> the nondominated set by solving 2P lexicographic programs. This is because different λ-vectors, even though they may be well dispersed, sometimes result in the same or very similar criterion vectors. But, by oversampling in Step 6, and then filtering away the most redundant, we should emerge with a group of nicely dispersed nondominated criterion vectors for presentation to the decision maker. Note that the 2P quantity is discretionary. It could be increased or decreased if more or less oversampling is desired.

<u>Step 8</u>: If h = t, terminate with $(z^{(h)}, x^{(h)})$ as the final solution, where $x^{(h)}$ is the inverse image of $z^{(h)}$ produced by the lexicographic sampling program that yielded $z^{(h)}$.

<u>Step 9</u>: Compute $\lambda^{(h)}$, the <u>unbiased λ-vector characterization</u> of $z^{(h)}$. The components of $\lambda^{(h)}$ are given by

$$\lambda_i^{(h)} = \frac{1}{(z_i^{**} - z_i^{(h)})} \left[\sum_{j=1}^{k} \frac{1}{(z_j^{**} - z_j^{(h)})} \right]^{-1}$$

This step is included because $z^{(h)}$ may not have been generated by the vertex of the intersecting contour of the lexicographic sampling program. When this is the case, it is better to use $\lambda^{(h)}$, because $\lambda^{(h)}$ is the λ-vector that would cause the lexicographic sampling program to uniquely generate $z^{(h)}$ at the vertex of an intersecting contour.

<u>Step 10</u>: Form $\Lambda^{(h+1)}$ so that its volume is r of the volume of $\Lambda^{(h)}$, where

(a) $\Lambda^{(h+1)} = \{\lambda \in \mathbb{R}^k \mid \lambda_i \in (\ell_i^{(h+1)}, \mu_i^{(h+1)}), \sum_{i=1}^{k} \lambda_i = 1\}$

(b) $\mu_i^{(h+1)} - \ell_i^{(h+1)} = 2w$ for all i where w is the <u>interval</u> <u>half-width</u>

$$(c)\quad (\ell_i^{(h+1)}, \mu_i^{(h+1)}) = \begin{cases} (0, 2w) & \text{if } \lambda_i^{(h)} - w \leq 0 \\ (1 - 2w, 1) & \text{if } \lambda_i^{(h)} + w \geq 1 \\ (\lambda_i^{(h)} - w, \lambda_i^{(h)} + w) & \text{otherwise} \end{cases}$$

<u>Step 11</u>: Go to Step 4.

The interval half-width w of Step 10(b) is computed by using the GRIDPOINT computer code developed by Liou (1984). By specifing $\lambda^{(h)}$ and r, GRIDPOINT is able to compute the $(\ell_i^{(h+1)}, \mu_i^{(h+1)})$ intervals of Step 10(c) such that

$$\frac{\text{Volume of } \Lambda^{(h+1)}}{\text{Volume of } \Lambda^{(h)}} = r$$

Recall that r, where $r < 1$, is externally specified in Step 1.

5. MODIFICATIONS

The basic algorithm of Section 4 can be modified depending upon the needs of the problem or the decision making style of the user. To illustrate, consider the five following modifications:

(1) <u>No Pre-chosens</u>. Suppose we do not want to commit to pre-chosen values for P, t and r. Rather, we would like to make decisions about P, r, and when to terminate, iteration by iteration. This can be accommodated in the algorithmic specification of Section 4 with the following deleted, inserted, and replaced steps.

<u>Step 1</u>: Delete.

<u>Step 4.5</u>: Specify the number of solutions P to be presented this iteration.

<u>Step 8</u>: If the user wishes to stop, terminate with $(z^{(h)}, x^{(h)})$ as the final solution, where $x^{(h)}$ is the inverse image of $z^{(h)}$ produced by the lexicographic sampling program that yielded $z^{(h)}$.

(2) Model Redefinition. Suppose the user would like to make alterations to the model in response to what is learned about the problem during the iteractive process. This can be accommodated in the algorithmic specification with the following inserted step.

Step 7.5: Modify the model and then recompute a z^{**} utopian criterion vector.

It is necessary to recompute z^{**} because the nondominated set may have changed. Note that if the modifications to the model are more than minor, an algorithm re-start is advised.

(3) e-Constraints. Suppose the decision maker would like to manipulate e-constraints after the first iteration. This can be accomplished with the following modified and inserted steps.

Step 3: Construct a z^{**} utopian criterion vector and let index set $G = \phi$.

Step 6: Once for each of the 2P dispersed λ-vectors, solve the lexicographic sampling program

$$\begin{aligned} \text{lex min } & \{\alpha, \sum_{i=1}^{k} (z_i^{**} - z_i)\} & \\ \text{s.t. } & \alpha \geq \lambda_i (z_i^{**} - z_i) & i \in K \\ & f(x) = z_i & i \in K \\ & z_i \geq e_i & i \in G \\ & x \in S & \\ & 0 \leqq \alpha \leqq \mathbb{R} & \end{aligned}$$

Step 8.5: Specify the index set G and an e_i for each $i \in G$.

(4) Augumented Sampling Program. If a lexicographic capability is not available for solving the lexicographic sampling program, the lexicographic sampling program can be replaced by the augmented sampling program. This can be accomplished with the following replaced step:

Step 6: Once for each of the 2P dispersed λ-vectors, solve the augmented sampling program

$$\min \{\alpha + \rho \sum_{i=1}^{k} (z_i^{**} - z_i)\}$$

$$\text{s.t.} \quad \alpha \geq \lambda_i(z_i^{**} - z_i) \qquad i \in K$$

$$f_i(x) = z_i \qquad i \in K$$

$$x \in S$$

$$0 \leq \alpha \in \mathbb{R}$$

where ρ is a sufficiently small, numerically significant, positive scalar.

Note that it is possible to encounter difficulties such as described in Steuer (1986, pp. 429-420) if ρ is not set small enough. In practice, however, values for ρ of .001 or .0001 can normally be counted on to function reliably.

(5) Re-scaling the Objectives. The sampling of the nondominated set works in the most unbiased fashion if the objectives are such that the ranges of the criterion values over the nondominated set for the different objectives are roughly the same. If this is not the case, the objectives should be re-scaled. Such re-scaling can be accomplished by using the π_i factors of (3.1) in the following revised step.

Step 3: Construct a payoff table. Re-scale the objectives using the π_i. Recomputer the payoff table. Form a z^{**} utopian criterion vector from the re-computed payoff table.

Note that the re-scaling is only done for internal purposes. Externally, the re-scaling is probably useless. Thus, everytime a criterion value crosses the interface from man to machine, it has to be multiplied by its π_i, and everytime a criterion value crosses the interface from machine to man, it has to be divided by its π_i.

6. SOME ADDITIONAL COMMENTS

A strength of the Tchebycheff procedure is that it can converge to non-extreme final solutions in multiple objective linear programming, and to unsupported final solutions in multiple objective integer and nonlinear programming. Also, the dispersed sampling of the nondominated set lends to the procedure a multiple starting point property which is helpful should the decision maker's utility function be nonconcave.

Like most other interactive procedures, the Tchebycheff procedure is applicable to large-scale programs when able to call conventional mathematical programming software to solve the optimization problems in Step 6. MINOS by Murtagh and Saunders (1977) is an example of the best type of code to use because it comes with its source. This is important because it is often desirable to make adjustments to the source for the most efficient implementation. A numerical illustration of the interactive Tchebycheff procedure is given in Steuer (1986, pp. 468-472), while applications of the Tchebycheff method are reported in Steuer (1983a), Steuer and Wood (1986), and Silverman, Steuer and Whisman (1988).

7. HOW THIS DESCRIPTION DIFFERS FROM PREVIOUS DESCRIPTIONS

This description of the Tchebycheff method differs from the descriptions in Steuer and Choo (1983) and Steuer (1986) in two ways. First, the lexicographic sampling program is stressed rather than the augmented sampling program. Second, the z^{**} utopian criterion vector is constructed to be strictly greater than z^*. This has two benefits. One is that we can now work with

$$\Lambda^{(h)} \subset \Lambda = \{\lambda \in \mathbb{R}^k |\ \lambda_i \in (0,1),\ \sum_{i=1}^{k} \lambda_i = 1\}$$

rather than with $\Lambda^{(h)} \subset \bar{\Lambda} = \{\lambda \in \mathbb{R}^k |\ \lambda_i \in [0,1],\ \sum_{i=1}^{k} \lambda_i = 1\}$. The other is that we avoid the concept of a definition point (Steuer (1986, p. 425)), substantially simplifying the presentation of the method. In addition, this paper shows how modifications such as in Section 5 can be integrated into the algorithmic specification of the procedure.

REFERENCES

[1] Benayoun, R., J. de Montgolfier, J. Tergny, and O. Larichev (1971). "Linear Programming with Multiple Objective Functions: Step Method (STEM)," Mathematical Programming, Vol. 1, No. 3, pp. 366-375.

[2] Dessouky, M. I., M. Ghiassi, and W. J. Davis (1986). "Estimates of the Minimum Nondominated Criterion Values in Multiple-Criteria Decision Making," Engineering Costs and Production Economics, Vol. 10, No. 2, pp. 95-104.

[3] Isermann, H. and R. E. Steuer (1988). "Computational Experience Concerning Payoff Tables and Minimum Criterion Values Over the Efficient Set," European Journal of Operational Research, Vol. 33, No. 1, pp. 91-97.

[4] Korhonen, P. and J. Laakso (1986). "A Visual Interactive Method for Solving the Multiple Criteria Problem," European Journal of Operational Research, Vol. 24, No. 2, pp. 277-287.

[5] Liou, F. H. (1984). "A Routine for Generating Grid Point Defined Weighting Vectors," Masters Thesis, Department of Management Science & Information Technology, University of Georgia, Athens, Georgia, 80 pp.

[6] Murtagh, B. A. and M. A. Saunders (1977). "MINOS User's Guide," Technical Report 77-9, Department of Operations Research, Stanford University, Stanford, California, 127 pp.

[7] Silverman, J., R. E. Steuer, and A. W. Whisman (1988). "A Multi-period, Multiple Criteria Optimization System for Manpower Planning," European Journal of Operational Research, Vol. 34, No. 2, pp. 160-170.

[8] Steuer, R. E. (1983a). "Multiple Criterion Function Goal Programming Applied to Managerial Compensation Planning," Computers and Operations Research, Vol. 10, No. 4, pp. 299-309.

[9] Steuer, R. E. (1983b). "Operating Manual for the ADBASE Multiple Objective Linear Programming Package," Department of Management Science & Information Technology, University of Georgia, Athens, Georgia, 102 pp.

[10] Steuer, R. E. (1986). Multiple Criteria Optimization: Theory, Computation, and Application, New York: John Wiley & Sons, 542 pp.

[11] Steuer, R. E. and E.-U. Choo, (1983). "An Interactive Weighted Tchebycheff Procedure for Multiple Objective Programming," Mathematical Programming, Vol. 26, No. 1, pp. 326-344.

[12] Steuer, R. E. and E. F. Wood (1986). "On the 0-1 Implementation of the Tchebycheff Solution Approach: A Water Quality Illustration," Large Scale Systems, Vol. 10, No. 1, pp. 243-255.

[13] Weistroffer, H. R. (1985). "Careful Usage of Pessimistic Values is Needed in Multiple Objectives Optimization," Operations Research Letters, Vol. 4, No. 1, pp. 23-26.

[14] Wierzbicki, A. P. (1980). "The Use of Reference Objectives in Multiobjective Optimization," Lecture Notes in Economics and Mathematical Systems, Vol. 177, Berlin: Springer-Verlag, pp. 468-486.

[15] Wierzbicki, A. P. (1986). "On the Complcteness and Constructiveness of Parametric Characterizations to Vector Optimization Problems," OR Spektrum, Vol. 8, No. 2, pp. 73-88.

[16] Zionts, S. and J. Wallenius (1976). "An Interactive Programming Method for Solving the Multiple Criteria Problem," Management Science, Vol. 22, No. 6, pp. 652-663.

[17] Zionts, S. and J. Wallenius (1983). "An Interactive Multiple Objective Linear Programming Method for a Class of Underlying Nonlinear Utility Functions," Management Science, Vol. 29, No. 5, pp. 519-529.

V I G -
A VISUAL AND DYNAMIC DECISION SUPPORT SYSTEM FOR MULTIPLE OBJECTIVE LINEAR PROGRAMMING

Pekka Korhonen * and Jyrki Wallenius **

* Helsinki School of Economics, Runeberginkatu 14-16, 00100 Helsinki, Finland

** University of Jyväskylä, Seminaarinkatu 15, 40100 Jyväskylä, Finland

Abstract

In this paper we describe the principles of VIG (Visual Interactive Goal Programming), a Multiple Criteria Decision Support System, recently developed by Korhonen. PARETO RACE is a corner-stone of this system, which is designed to support both the modelling and solving of a multiple objective linear programming problem. The interface is based on one main menu, spreadsheets, and interactive use of computer graphics. VIG provides the decision-maker with the possibility to approach his/her decision problem by using an "evolutionary approach". This means that the decision-maker does not have to specify the model precisely prior to solving the problem. In fact, the model evolves progressively. We also discuss several applications of VIG to practical problems.

Keywords: Decision Support, Computer Graphics, Multiple Criteria, Evolutionary Approach.

1. INTRODUCTION

As Weber [1986] points out, the concept of Decision Support Systems (DSS) is now more than a decade old. Decision Support Systems have over the years been developed and successfully implemented in many organizations by individuals at different organizational levels. Various definitions have been offerred for decision support. Ginzberg and Stohr [1982] define a DSS as "a computer-based information system used to support decision-

NATO ASI Series, Vol. F56
Multiple Criteria Decision Making and Risk Analysis
Using Microcomputers. Edited by B. Karpak and S. Zionts

making activities in situations where it is not possible or not desirable to have an automated system perform the entire decision process". Multiple Criteria Decision Support Systems (MCDSS) are considered by Jelassi et al. [1985] as a "specific" type of system within the broad family of DSS. Even though MCDSS include much the same components as "traditional" Decision Support Systems, MCDSS have special characteristics that distinguish them from other DSS. Such characteristics include the fact that they allow analysis of multiple criteria; they use a variety of multiple criteria decision methods to compute efficient solutions; they incorporate user's input in various phases of modelling and solving a problem. While it has been customary to consider algorithms as the focal point of decision support, emphasis is shifting to the database and modelling activities. See, for example, Keen and Scott-Morton [1978], Sprague and Carlson [1982], Bui [1984], and Van Hee [1987].

In this paper we describe the principles of VIG (Visual Interactive Goal Programming), a Multiple Criteria Decision Support System recently developed by Korhonen [1987a]. The system is user-friendly and it is designed to support both the modelling and solving of a multiple objective linear programming problem. Menus, spreadsheets, and interactive use of computer graphics play a central role. Constraints and objectives are treated uniformly; their role may be changed during a session. Such features make it possible to apply an "evolutionary approach" to solving linear decision models (Korhonen, Narula, and Wallenius, 1988). In this approach, the decision-maker (DM) does not have to specify the model precisely prior to solving the problem. In fact, the model evolves progressively. The search for the most preferred solution takes place on the efficient frontier specified by the set of current objectives.

VIG is written in TURBO PASCAL and implemented on an IBM PC/1 microcomputer. To benefit fully from VIG, a color monitor should be used. In the system, PARETO RACE, a visual, dynamic, search procedure for exploring the efficient frontier of a multiple objective linear programming problem, plays a central role (Korhonen and Wallenius, 1988). In PARETO RACE the user sees the objective function values (flexible goals) on a display in numeric form and as bar graphs, as he travels along the efficient frontier. The keyboard controls include an accelerator, gears, brakes, and a steering mechanism.

This paper consists of seven sections. The introduction explains the underlying philosophy of VIG. The second section provides the problem formulation. The third section describes the structure of VIG. The fourth section discusses PARETO RACE. The fifth section describes the "evolutionary approach". In the sixth section several applications are described. The seventh section concludes the paper.

2. PROBLEM FORMULATION

Let us consider the following problem, where the consequences (outcomes) y_i, $i = 1, \ldots, m$, of decisions (activities, actions, choices) can be stated as linear functions of the decision variables x_j, $j=1,\ldots,n$:

$$y_i = y_i(x) = \sum_{j=1}^{n} a_{ij}x_j \ , \ i \in M = \{1,2,\ldots,m\},$$

or equivalently in the matrix form:

$$\mathbf{y} = \mathbf{y}(\mathbf{x}) = A\mathbf{x},$$

where $\mathbf{x}$ is an n-vector of decision variables, A is an m x n matrix of coefficients, and $\mathbf{y}$ is an m-vector of consequences or outcome variables. In the subsequent discussion, the vector $\mathbf{y}$ may include the vector $\mathbf{x}$, if the DM imposes restrictions upon it or has preferences about the values of the decision variables. The problem is to find values for the decision variables x_j, $j \in N = \{1,2,\ldots,n\}$, such that the outcome variables, y_i, $i \in M$, would have acceptable or desirable values. If $n \geq m$, then for each desired or given value of $\mathbf{y}$, there exist an infinite number of solutions for the model, and it can easily be solved. To avoid this trivial case, we assume that $m > n$.

Various procedures have been developed for solving the problem. These include linear programming, fuzzy linear programming, multiple objective linear programming, and "what-if" -analysis. Each of these approaches imposes certain limitations on the model that may result in misrepresenting real-world decision situations.

In linear programming, one of the outcome variables is chosen as an objective function to be maximized or minimized over a feasible set. The feasible set is defined by specifying acceptable values for the other outcome variables. As is well known, it is possible to find an optimal solution (if one exists) for a linear programming problem. Implicitly, in linear programming we assume that the DM's preference structure (or the value function) over the feasible set can be adequately described by means of a single objective function. Furthermore, although any boundary point is acceptable, a point that lies just outside the feasible region is not. Thus, in linear programming there is a very clear conceptual difference between an objective function and the constraints. Duality theory and postoptimality analyses can be used to study the neighborhood of the "optimal" solution, but such techniques may not provide enough flexibility for the DM.

In fuzzy linear programming, the DM is not assumed to minimize or maximize an objective function (Zimmermann, 1985). Instead, he/she is assumed to want to reach some "fuzzy" aspiration levels. In fuzzy linear programming, membership functions are used to describe the violation of constraints. Each membership function has value 1, if the constraint is fully satisfied, and 0, if the maximum tolerance is exceeded. The simplest type of membership functions are linear, but several other forms of membership functions and aggregation rules are also possible (see, for example, Sakawa, 1983). However, after one has specified such functions and aggregation rules, the solution procedure is straightforward and a unique solution is found without the DM's intervention. Some methods have recently been developed, which allow the DM to analyze the effect of "fuzziness" on the final solution (Chanas, 1983, Carlsson and Korhonen, 1986). To sum up, usually in fuzzy linear programming, the boundary of the feasible region is "soft", whereas the final solution is "hard".

In multiple objective linear programming, the DM selects a number of outcome variables as objective functions and attempts to maximize (or minimize) them, simultaneously. Since this problem has rarely, if ever, a unique solution, the DM has to choose a solution from among the set of nondominated (efficient) solutions (see, Steuer, 1986). As in linear programming, the DM has to make a sharp distinction between the constraints and the objective functions at the beginning of the solution process. Also, all preference information is assumed to be contained in the objective functions. However, sometimes seeing the solution makes the DM realize that it is not satisfactory. Thus, despite the advanced state of research, even a multiple objective approach may not always be flexible enough in practice.

The DM may also experience difficulties in specifying, what he/she wants and what it is possible to achieve. Using "what-if" analysis, the DM can experiment with different values of the decision variables and consider the values of the outcome variables without specifying explicit bounds for them. Alternate solutions are generated until the DM is satisfied. However, "what-if"-analysis does not guarantee that the most preferred solution will be found. In fact, the final solution may be dominated.

To overcome some of the above difficulties, Korhonen et al. [1988] have recently proposed that an evolutionary approach be used to solving the problem. Briefly, in the evolutionary approach the modelling and solving phases are not separated as is usually done, but instead are considered simultaneously. Also, there is initially no conceptual difference between objective functions and constraints. The evolutionary approach may (and in fact has been) implemented through VIG.

In the evolutionary approach, the problem is formulated as follows (see, Korhonen and Wallenius, 1987):

$$\min \epsilon$$

$$\text{subject to:} \qquad (2.1)$$

$$\begin{aligned} C\mathbf{x} + \mathbf{w}\epsilon &\geq \mathbf{b}, \\ \mathbf{x} &\geq \mathbf{0}, \end{aligned}$$

where **w** is an m-vector whose components are

$$w_i \quad \begin{cases} =0, \text{ if } i \text{ refers to a constraint,} \\ >0, \text{ if } i \text{ refers to an objective.} \end{cases}$$

C is a k x n matrix of coefficients, **b** is an m-vector of aspiration levels, and ϵ is a scalar variable. At the optimum of ϵ, the solution vector **x** is (at least) weakly efficient. Note that using the notation in section 1:

$$A = \begin{bmatrix} C \\ I \end{bmatrix}.$$

To move about on the efficient frontier, we can vary either the vector **w** or the aspiration levels **b**. We adopt the latter strategy. If only the components of **b** corresponding to the objective function rows are varied, then we move on the efficient frontier and the constraint set remains unchanged. Of course, we can also change the components of **b** corresponding to the other rows, thus changing the constraint set. Actually, then we have a classical parametric programming problem. Note that the values of the **w**-vector specify, which rows are objectives and which constraints. By making changes in **w**, we can easily change the role of objectives and constraints.

3. THE STRUCTURE OF THE SYSTEM

VIG is implemented in the spirit of visual interaction. Following the spirit of goal programming, constraints are regarded as a subset of goals (Ignizio, 1983). In fact, constraints are inflexible goals. VIG is described in more detail in the user's guide (Korhonen, 1987b).

To illustrate the use of the system, we consider a production planning problem, where the DM tries to find the "best" product-mix for three products: Product 1, Product 2, and

Product 3. The production of these products requires the use of one machine (Mach.Hours), man-power (Man Hours), and two critical materials (Crit.Mat 1 and Crit.Mat 2). Selling the products results in profit. The A-matrix of the problem is given in Table 1:

	Product 1	Product 2	Product 3
Mach.Hours	1.5	1.0	1.6
Man Hours	1.0	2.0	1.0
Crit.Mat 1	9.0	19.5	7.5
Crit.Mat 2	7.0	20.0	9.0
Profit	4.0	5.0	3.0
Product 1	1.0	0.0	0.0
Product 2	0.0	1.0	0.0
Product 3	0.0	0.0	1.0

TABLE 1: The Coefficient Matrix of the Model

Suppose that the DM has difficulties in clearly specifying, what he wants to achieve and what the bounds for the (inflexible) goals might be. For example, the DM may state:

"I would like to make as much profit as possible, but because it is difficult to obtain critical materials, I would like to use them as little as possible. One machine is used to produce the products. The machine operates without any problems for at least 9 hours, but it is very likely to break down if it is used for more than 12 hours. The length of a normal working day is 10 hours. People are willing to work overtime, but it is costly and they are tired the next day. Therefore, if possible, I would like to avoid it. Finally, product 3 is very important to a major customer, and I cannot totally exclude it from the production plan."

The preceding decision problem is typically semi-structured. In the DSS literature (see, e.g., Sprague and Carlson, 1982), "what-if" analysis has been proposed to solving such problems. However, the "what-if" -approach does not provide enough support for finding the best solution.

Throughout this paper we use this example to illustrate the use of our system.

VIG has four main functions:

(1) **Model Management Function**

(Allowing us to create, edit, store, and retrieve models.)

(2) **Model Development Function**

(Allowing us to provide the decision variables and outcome variables (goals) with names, specify the technolgy matrix describing the relationships among the variables, and determine aspiration levels for the goals.)

(3) **Problem Solving Function**

(Allowing us to tell the system, how the problem should be solved - and actually to solve it. With more than two objectives (flexible goals), PARETO RACE is invoked. With one or no objectives, no user's intervention is necessary.)

(4) **Solution Output Function**

(Allowing us to examine solutions on the screen, and store intermediate results for later consideration or for writing reports.)

```
                              V I G
                             *******

               The Name of the Model:    <None>

                              Menu
                              ****

               ==>  Select Model
                     Name Rows and Columns
                     Edit Matrix Coefficients
                     Specify the Types and Values of Goals
                     Save the Model
                     Classify the Goals
                     Specify Ranges for Flexible Goals
                     Solve the Problem
                     Display the Values of Goals
                     Display the Values of Decision Variables
                     Save the Current Solution
                    END

<┘:Proceed F10:As <┘
```

Figure 1. The Main Menu

Each main function consists of one or more sub-functions, which are all chosen from the main menu. At each stage, the available choices are shown using light cyan. All functions are not available all the time. For example, one cannot examine a solution before it is computed. Using only one main menu makes it easy to keep the control of the program in the user's hands. The user is welcome to choose anyone of the available functions without "having to run through the entire list". When a function is terminated (using the F10-key), the system prints the main menu. By moving the cursor up and down, we can trace all available choices.

Model Management Function

"Select Model" and "Save the Model" activate Model Management Functions. When we enter "Select Model", the system prompts us to provide the model with a name. If we are creating a model, the system uses "Temp" as a default name. If we wish to save the model, another name should be used. "Temp" will be erased at the end of the session. By entering "Save the Model" we store the current version of the model into the default directory (diskette). We can rename it or use the old name.

Model Development Function

The model is entered in three phases. Each phase has its own spreadsheet format. Using arrow keys, we can move from one field to another. The use of any other key activates the field, and makes it ready for our input.

Initially, we are asked to provide the decision variables (columns) and goals (rows) with names. At this stage, we are urged to consider the question: "What are my options and what are their consequences?", without incorporating numbers. When creating a model, only the frames of the names are provided.

Next, we enter "Edit Matrix Coefficients" and specify the technology matrix of the problem. In other words, we specify the dependences of the goals on the decision variables. At this stage, the relevant question is: "How will my decisions affect the values of the goals?"

Editing the Types and Aspiration Levels of Goals

Names	Types	Given Values	Current Values
Mach.Hours	<=	9.0000000	0.0000000
Man Hours	<=	10.000000	0.0000000
Crit.Mat 1	<=	96.000000	0.0000000
Crit.Mat 2	<=	96.000000	0.0000000
Product 1	>=	0.0000000	0.0000000
Product 2	>=	0.0000000	0.0000000
Product 3	>=	0.0000000	0.0000000
Profit	>=	28.000000	0.0000000

F10:Exit to Menu

Figure 2. Aspiration Levels

Finally, we define aspiration levels for the goals (the right-hand-side vector)(Fig. 2: Given Values). We also define the directions of the inequalities (Types). The direction indicates, whether we wish to exceed the aspiration levels or fall below them. For example, the "greater than or equal" symbol associated with Profit indicates that we wish to maximize the increase in Profit. Similarly, the "less than or equal" symbol of critical materials indicate minimization of the use of materials.

Note that up to this point we have not made any distinction between objectives and constraints. We have treated them in a uniform manner.

Problem Solving Function

This function consists of three phases: Classifying the Goals, Specifying Ranges for Flexible Goals, and Solving the Problem. Initially, all goals are assumed to be inflexible (that is, constraints). Moving the cursor to point to a row name and pressing SPACE BAR once changes the status of the goal (from inflexible to flexible, and vice versa). On the screen, brown stands for inflexible and white for flexible goals.

If none of the goals is specified to be flexible, the program will find a feasible solution, if one exists. If one of the goals is specified flexible, the program solves a linear programming problem after we have entered "Solve the Problem". This is done without user's intervention. If more than one goal is flexible, the system invokes PARETO RACE. PARETO RACE needs the user to specify approximate ranges for the flexible goals (Fig. 3). In our example, initially we have three flexible goals. Mach.Hours is (initially) allowed to vary from 8 to 13, Man Hour from 9 to 14, and Profit from 25 to 35.

Specifying (Roughly) the Bounds for Goals

Names	Types	Given Values	Current Values	Lower Bounds	Upper Bounds
Mach.Hours	<=	9.0000000	0.0000000	8.0000000	13.000000
Man Hours	<=	10.000000	0.0000000	9.0000000	14.000000
Profit	>=	28.000000	0.0000000	25.000000	35.000000

F10:Exit to Menu

Figure 3. Bounds for Flexible Goals

Ranges for flexible goals are needed for defining the lengths of the bars on the screen. To obtain a maximum benefit from the visual feature of PARETO RACE, we recommend careful consideration of the ranges (even though they are automatically updated by the system). Sometimes it is advisable to respecify the ranges during the race. In the use of the program, experience is a good teacher.

By entering "Solve the Problem", the PARETO RACE screen is displayed (Fig. 4). Since PARETO RACE is a corner-stone of our system, it is explained more in detail in the next section.

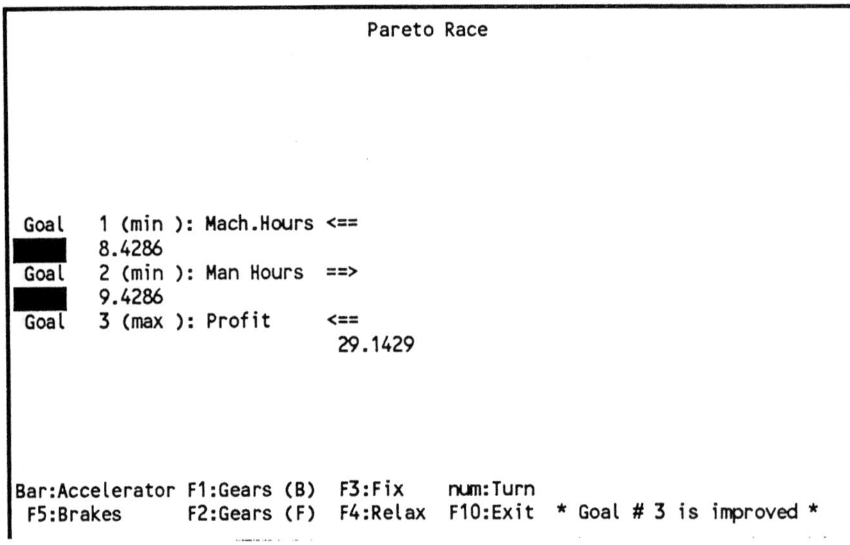

Figure 4. An Example of PARETO RACE: Initial Solution (Making a Turn)

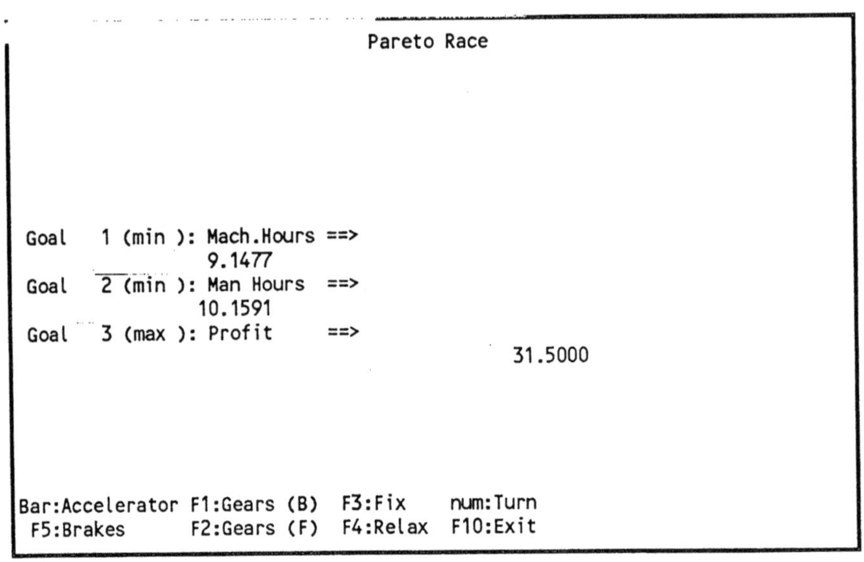

Figure 5. An Example of Another Solution of PARETO RACE

Solution Output Function

After using PARETO RACE, the choices "Display the Values of Goals", "Display the Values of Decision Variables", and "Save the Solution" become available. They enable us to examine the values of all flexible and inflexible goals, and the values of nonzero decision variables. We can save intermediate solutions on a diskette or a hard disc. It is possible to store all solutions in the same file. Afterwards we may examine solutions using a text processing system or a text editor.

Having obtained one solution, we can go back and edit the model, change the set of flexible goals, or respecify the ranges for the flexible goals for a better visual effect. We can also return to PARETO RACE. The race starts from the current solution, if we have not edited the model in the meantime.

4. PARETO RACE

In PARETO RACE, we can freely search the efficient frontier of a multiple objective linear programming problem. Specific keys are used to control the speed and direction of motion. On a display, we see the objective function values in numeric form and as bar graphs whose lengths are dynamically changing as we move about on the efficient frontier. The keyboard controls include the following function keys (see, Fig. 4 and Fig. 5):

(SPACE) **BAR:** An "Accelerator"

We proceed in the current direction at constant speed.

F1: "Gears (Backward)"

Increase speed in the backward direction.

F2: "Gears (Forward)"

Increase speed in the forward direction.

F3: "Fix the Aspiration Level of a Specific Goal"

The current value of a specific goal is taken as an absolute lower (upper) bound.

F4: "Relax the Aspiration Level of a Fixed Goal"

The fixed goal becomes flexible again.

F5: "Brakes"

Reduce speed.

num: "Turn"

Change the direction of motion by pressing the number key corresponding to the goal's ordinal number once or several times.

F10: "EXIT"

Exit to the main menu.

Initially, we are going to see (graphically and numerically) an efficient solution on the computer screen (Fig. 4). Arrows indicate an initial direction chosen by the computer (based on our aspiration levels). If we like this initial direction, we hold SPACE BAR down and observe the solutions change. We are "travelling" at base speed. If we like to increase the speed, we press the F2-key once or several times (depending on the desired speed) and hold SPACE BAR down. If at some point the direction is no longer attractive, we initiate a turn (see, Fig. 4). Assume that the user wishes to improve a certain goal. To accomplish this, the goal's corresponding number key is pressed once or several times, depending on the amount of desired improvement. Then, the program updates the direction, and so forth. To reduce speed, we use brakes (F5-key). If we wish to resume base speed, the most convenient way is to press the F1-key (gears: backward) and then the F2-key (gears: forward). We reverse direction twice and start travelling at base speed. When a flexible goal is fixed by using the F3-key, the system asks us to specify the number of the goal to be fixed; then an asterisk appears next to the name of such a goal. The goal may be relaxed by using the F4-key; the asterisk disappears.

When we terminate PARETO RACE, we are welcome to examine the values of the decision variables and goals. Just exit to the main menu. In fact, the user can examine the values of the decision variables and goals also during the race (and save such intermediate results for later use). Simply enter "Solve the Problem" in the main menu and we are back in the race. We can also change the role of flexible and rigid goals during the race. We feel that this adds flexibility to the system and extends it beyond "classical" multiple objective linear programming.

5. AN EVOLUTIONARY APPROACH

The evolutionary approach is explicitly based on the idea that there is no need to make a conceptual distinction between objective functions and constraints, and makes their treatment more systematic. The preference function (value function) u is defined for all outcome variables including the decision variables. We do not assume that the value function can be represented using a static, mathematical functional relationship. However, without loss of generality we assume that u is an increasing function of its arguments. Thus, each y_i, $i=1,2,...,m$, is assumed to be maximized.

In practice, since the number of outcome variables may be quite large, it may be impossible to consider all outcome variables as objectives, simultaneously. As an alternative, Korhonen et al. [1988] have proposed a systematic approach which allows the DM to consider any subset of outcome variables as objectives at a time.

A step by step statement of the approach follows:

0. Set t:=1.

1. Specify the index set G of the outcome variables y_i, $i \epsilon G \subset M$, which are to be treated as objective functions. Denote by R the index set of the remaining variables, R=M-G. Without loss of generality, assume that $G=\{1,2,...,s\}$. Now, denote $\mathbf{y}^1=(y_1, y_2, ...,y_s)$ and $\mathbf{y}^2=(y_{s+1}, y_{s+2}, ..., y_m)$.

2. Specify set Q using bounds for the variables $\mathbf{y}^2$ as follows:

$$Q = \{ \mathbf{y}^2 \mid b_i \leq y_i \leq c_i, \quad i \epsilon R\},$$

such that for any given $\mathbf{y}^1$, the marginal value function $u(\mathbf{y}^1,\mathbf{y}^2)$ does not "vary much" when $\mathbf{y}^2 \epsilon Q$.

3. Find the most preferred values for $\mathbf{y}^1$, subject to $\mathbf{y}^2 \epsilon Q$. Set t:=t+1 and denote the solution by y_t.

4. If the DM is satisfied with the solution y_t, stop; otherwise return to step 1.

It is possible to implement the proposed procedure using VIG. In the proposed procedure, the DM is an essential and integral part of the system, and he/she has complete freedom to change the bounds and the roles of the objectives and constraints during the process. In the above algorithm, the DM can go to or return to any step (except step 0) from any other step. This can be accomplished by using a menu-interface (see, Korhonen, 1987).

An essential feature in the evolutionary approach is that all the time we are moving on the efficient frontier. Of course, the efficient frontier changes when we change the set of objectives and constraints. However, the search on the current efficient frontier is conducted much more systematically than a search over a "softly-defined feasible set".

Let us consider our example again. Suppose we start with the profit maximization model (G = {5} in step 1), where set Q is defined using the same bounds as in the LP formulation. We find the solution of the LP-problem in column I of Table 2.

Names		**Types**	**Initial**	**Computed Values**		
		Values	**I**	**II**	**III**	**IV**
Mach. Hours	$\le$	9.0	9.0	10.0*	10.0	10.0*
Man Hours	$\le$	10.0	10.0	10.2*	10.2	10.4*
Crit. Mat 1	$\le$	96.0	94.5	96.0	94.3	96.0
Crit. Mat 2	$\le$	96.0	88.0	87.6	89.6	91.5
Profit	$\ge$	$-\infty$	31.0*	32.9*	31.7*	32.0*
Product 1	$\ge$	0.0	4.0	4.9	3.8	3.7
Product 2	$\ge$	0.0	3.0	2.7	2.7	2.8
Product 3	$\ge$	0.0	0.0	0.0	1.0*	1.0

TABLE 2: An Evolutionary Solution Procedure
("*" refers to the outcome variables used as objectives)

Now, when the DM realizes that machine hours and man hours are binding constraints, he may want to consider them and profit simultaneously as objectives. His value function does not change much, if he exceeds the bounds somewhat. He returns to step 1, redefines G = {1,2,5}, and uses the same bounds for the remaining variables as before. The efficient solution given under column II of Table 2 is our next solution. In this solution the machine is used one extra hour and the regular work time is exceeded by 0.2 hours. Now, the first critical material has become binding. However, the third product is not produced at all. The DM feels that he cannot increase the machine hours or man hours any further. The DM decides to consider profit and the third product as objectives for the next iteration (G={5,8} in step 1) and uses the current values of machine hours and man hours as upper bounds in step 2. An efficient solution is shown in column III of Table 2. Note that there is a sharp drop in profit, when one unit of the third product is produced. This may not be acceptable to the DM. Suppose that one unit of the third product is enough for the DM's needs, so he specifies one as a lower bound (step 2). Further, since machine hours and man hours are binding, he may want to reconsider them and profit as objectives (G={1,2,5} in step 1). Again, an efficient solution is provided in column IV of Table 2, and so forth. The process is terminated, when the DM has found a satisfactory solution.

6. APPLICATIONS

VIG has been and is being applied to solving several practical problems. In this section we briefly survey a number of such applications. Additional details may be found in the original references.

One of the earliest applications deals with the determination of optimal price changes for alcoholic beverages sold by ALKO Ltd., the Finnish State Alcohol Monopoly. Pricing decisions are among the most important alcohol policy measures in Finland. The problem is not only to maximize profits, but at the same time restrict sales with the intention of reducing harmful effects due to alcohol consumption, and to minimize the impact on the consumer price index. A multiple objective linear programming model with three objective functions (reflecting the above concerns) was developed and solved using our procedure. The decision variables of the model were logarithms of relative price changes. Modifications in the input-output routines of the program enabled the user to operate with original percentages instead of their logarithms. Based on this model, the company has implemented an extended version of the model. The ultimate decision-maker, the governmental ministers comprising the cabinet, is interested in the model. Additional details may be found in Korhonen and Soismaa [1988].

Another interesting and important application concerns the problem of stockpiling critical materials for a national emergency in Finland. It is conceptually a bi-criteria problem.

On the one hand, the purpose is to minimize the cost of stockpiling, and on the other hand to maximize the welfare of the population at large during (and after) a crisis. Actually, the welfare of the population is measured using four criteria: the consumption rate in each of the four quarters. Thus, the model is considered as a five criteria form. Using the model it is possible to make intelligent decisions regarding appropriate stockpiling levels for critical materials, and understand the implications in case of a crisis. The model is being used by the National Board of Economic Defense. No scientific article has been published on the subject, but the system and its merits have been summarized in some Finnish newspapers.

In Kananen et al. [1988] we demonstrate how VIG can effectively be used for analyzing input-output models. We have applied our approach to studying the quantitative effects of economic or political crises to the Finnish economy. Examples of such crises are nuclear power plant accidents, trade embargoes, and international conflicts of various nature. An input-output model, based on the latest official input-output statistics of the Finnish economy with 17 industries (sectors) is employed. Relevant objective functions depend on the planning problem, but most frequently used are maximization of (private) consumption, employment, and total gross outlay. Our system is currently being used by the National Board of Economic Defense. For large-scale problems, such as inter-temporal, regional, input-output models, a large-scale version of VIG should be used.

An interesting, prototypical application of VIG (called

marginal planning) has been performed at a major Finnish wholesale company. The project somewhat resembles our input-output study of the Finnish economy. Yet, this problem concerns a division of a company split into 11 departments (units). The analysis starts from the income statements of each department. Each of these income statements forms a column in our model. Some of the items (rows) in the income statements (for example, profits, turnover, etc.) are defined as objective functions (flexible goals). Using VIG the company management can decide what they want to achieve (division-wise) in terms of profits, turnover, and so forth. Then, they can find out, which departments are instrumental (critical) in achieving these overall goals, and which departments need strengthening in terms of managerial skills, capital, labor. Interestingly, our system pinpointed certain weak departments in the division (that were known to the management) plus revealed some additional information. The objectives may be changed in the spirit of the evolutionary approach. Based on this study, the company reached a decision to start using VIG at management level. The results will be forthcoming as future reports.

We have applied VIG to solving the problem of advertising media selection in a Finnish software company. The problem was basically to assist the management in allocating an advertising budget across various media. The purpose was to maximize audience exposure. Six different newspapers and professional magazines were initially chosen as the relevant advertising media. The relevant audience consisted of the following (target) groups:

Marketing Management, Finance and Personnel, ADP, Production Management, R&D, and General Management. To measure the audience exposure, readership numbers were obtained from Finnish Gallup Ltd., reflecting the current situation. The media selection problem was modelled as a linear decision problem in the spirit of the evolutionary approach. The decision variables were the numbers of ads in each medium, which were also considered as consequences in our model. For additional details, an interested reader may consult Korhonen et al. [1988].

The waters of the New York Bight are among the most intensively utilized in the world. Concern for water quality in this region is long standing. Yet, sewage sludge has been dumped at the so-called 12-mile site for more than 60 years. Now the communities using the site are shifting their operations to the more distant 106-mile site, following orders issued in 1985 by the U.S. Environmental Protection Agency. In a recent study, Wallenius et al. [1987] use VIG to re-examine the EPA decision in a way which permits simultaneous multi-site dumping. The study is based on a multiple objective linear programming model, providing a framework for ocean waste disposal management. The relevant objective functions were to minimize transportation costs to New York City, and communities in New Jersey and on Long Island, and to minimize the volume of pollutants in near-shore and off-shore areas. The decision variables in the model are the number of barge trips made from source i to site j by disposal method k. The model consists of 40 decision variables and 27 constraints (inflexible goals) imposing environmental and capacity restrictions.

Duckstein et al. [1988] describe the use of VIG in forest management. More specifically, they examine the case of managing a ponderosa pine forest in the Salt-Verde river basin, situated in North-Central Arizona. The problem is to allocate to given subwatersheds six different treatment strategies corresponding to clearcutting between 0 and 100% of the area. This allocation is evaluated using multiple, conflicting objective functions, such as water yield, forage production, sediment yield, amount of recreational usage, economic benefits/costs. The problem has a number of physical constraints, some of which are fuzzy.

Last but not least, we would like to mention the work done by Joen Forest Consulting, a Finnish consulting company specialized in forest sector models. VIG is an important part of their system for forest planning that will be used by more than half a dozen forest sector (vocational) educational institutions in Finland.

7. CONCLUSION

In this paper we have described a software package called VIG. It is essentially a Multiple Criteria Decision Support System for modelling and solving linear programming problems. The computer program is written in TURBO PASCAL and it implements PARETO RACE as an essential ingredient. The necessary hardware consists of an IBM compatible Personal Computer with at least one diskette drive and a graphics card. The minimum memory size is 256 Kbytes. Preferably, a color monitor should be used. The current version of the program is capable of solving problems with a maximum of 96 variables and 100 goals, from which at most ten may be flexible, simultaneously. The size of the array may be increased, although part of the dynamic effect is then lost. The program consists of more than 3000 lines of code, most of which is used to build up an attractive interface. One can easily construct or edit one's models with computer graphics playing a central role. VIG was updated in August of 1987, with many desirable features added.

Our objective is to make the program widely available, and pursue several applications.

References

Bui, X. T. (1984), "Building Effective Multiple Criteria Decision Support Systems", Systems, Objectives, Solutions 4, Nr. 1, 3-16.

Chanas, S. (1983), "The Use of Parametric Programming in Fuzzy Linear Programming", Fuzzy Sets and Systems, 11, 243-251.

Carlsson, C. and Korhonen, P. (1986), "A Parametric Approach to Fuzzy Linear Programming", Fuzzy Sets and Systems, 20, pp. 17-30.

Duckstein, L., Korhonen, P. and Tecle A. (1988), "Multi-objective Forest Management: A Visual, Interactive and Fuzzy Approach", (The paper presented at the Conference of Systems Analysis in Forest Resources Management: The 1988 Symposium, Asilomar, CA, 30 March - 1 April, 1988).

Ginzberg, M. and Stohr, E. (1982), "Decision Support Systems: Issues and Perspectives", in: M. Ginzberg, W. Reitman, W., and E. Stohr, Eds., Decision Support Systems, (North-Holland Publ. Comp., Amsterdam, 1982).

Ignizio, J. P. (1983), "Generalized Goal Programming", Computers and Operations Research 10, (1983) 277-289.

Jelassi, M. T., Jarke, M. and Stohr, E. (1985), "Designing a Generalized Multiple Criteria Decision Support System", Journal of Management Information Systems I, (1985) 24-43.

Kananen, I., Korhonen, P., Wallenius, H. and Wallenius, J. (1988), "A Multiple Objective Approach to Analyzing Iput-Output Models, with an Application to Economic Defense", (unpublished manuscript).

Keen, P. G. W. and Scott-Morton, M. (1978), Decision Support Systems: An Organizational Perspective, Addison Wesley.

Korhonen, P. (1987b), "VIG - A Visual Interactive Support System for Multiple Criteria Decision Making", Belgian Journal of Operations Research, Statistics and Computer Science 27, Nr. 1, 3-15.

Korhonen, P. (1987b), VIG (A Visual Interactive Approach to Goal Programming) -User's Guide.

Korhonen, P., Narula, S., and Wallenius, J. (1988), "An Evolutionary Approach to Decision-making, with an Application to Media Selection", (unpublished manuscript).

Korhonen, P. and Soismaa, M. (1988), "A Multiple Criteria Model for Pricing Alcoholic Beverages", (Forthcoming in European Journal of Operational Research).

Korhonen, P. and Wallenius, J. (1986), "A Pareto Race", Working Paper (DIS 85/86 - 13), Arizona State University, (Revised in August 1987), (Forthcoming in Naval Research Logistics, in 1988).

Sakawa, M. (1983), "Interactive Computer Programs for Fuzzy Linear Programming with Multiple Objectives", International Journal of Man - Machine Studies, 18, pp. 489-503.

Sprague, R. H. and Carlson, E. C. (1982), Building Effective Decision Support Systems, Prentice-Hall.

Steuer, R. (1986), Multiple Criteria Optimization: Theory, Computation, and Application, John Wiley & Sons, New York.

Van Hee, K. M. (1987), "Features of the Architecture of Decision Support Systems", Paper presented at the 12th Symposium on Operations Research, Gesellschaft fur Mathematik, Ökonomie und Operations Research, Universität Passau, September 9-11.

Wallenius, H., Leschine, T. M. and Verdini, W. (1987), "Multiple Criteria Decision Methods in Formulating Marine Pollution Policy: A Comparative Investigation", unpublished manuscript.

Weber, E. S. (1986), "Systems to Think With: A Response to "A Vision for Decision Support Systems"", Journal of Management Information Systems II, 85-97.

Zimmermann, H.-J. (1976), "Description and Optimization of Fuzzy Systems", International Journal of General Systems, 2, pp. 209-215.

Zimmermann, H.-J. (1985), Fuzzy Set Theory - and Its Applications, Kluwer-Nijhoff Publishing, Boston.

A VARIATION OF THE METHODS OF ZIONTS AND WALLENIUS

Murat M. Koksalan
Cagla Beseli
Department of Industrial Engineering
Middle East Technical University
Ankara 06531 TURKEY

ABSTRACT

The type and amount of information required of the decision maker are among the main concerns in multiple criteria decision making. In this paper we develop a method that requires a reasonable amount of relevant information of the decision maker for multiobjective linear programming. The method is a combination of the methods of Zionts and Wallenius [1976,1983]. We compare the method with that of Zionts and Wallenius [1983] on a number of problems.

1. INTRODUCTION

The problem we consider is to maximize an unknown utility function of p linear objectives over a convex region. The motivation to this study are the methods of Zionts and Wallenius [1976,1983].

Multiple criteria decision making methods usually require some type of preference information of the decision maker. Using the information, methods try to find the most preferred solution, several preferred solutions or rank order the solutions. The type and amount of required information are

NATO ASI Series, Vol. F56
Multiple Criteria Decision Making and Risk Analysis
Using Microcomputers. Edited by B. Karpak and S. Zionts

important factors in reaching preferred solutions. It may be hard for the decision maker (DM) to respond to certain type of preference questions and as a result the solution(s) suggested by the method may be poor solutions. Similarly, the DM may not be willing to provide a large amount of preference information.

The method of Zionts and Wallenius [1976] require tradeoff questions which are usually not very easy for the DM to respond. They require the DM to answer tradeoff questions as well as to choose between pairs of distinct solutions in their more recent method [1983]. Individuals, in general, do not have difficulty in comparing distinct solutions when the number of criteria is not too large (see Miller [1956] and Green and Srinivasan [1978]).

In this study we develop a method, which is mainly a mixture of the above methods of Zionts and Wallenius. The method requires the DM only to compare pairs of distinct solutions. We compare the method with that of Zionts and Wallenius [1983] on a number of problems on the basis of amount of required information and choosing a highly preferred solution.

2. OVERVIEW OF THE ZIONTS AND WALLENIUS (Z-W) METHODS

Z-W [1976] method tries to identify a solution such that the DM does not like any tradeoff leading to adjacent efficient solutions. The method uses the responses of the DM in order to restrict the region which defines the possible values of weights of a linear utility function. The utility function, with updated weights, is used at each iteration to find a maximizing solution.

The maximizing solution is the basis for the next iteration.

Z-W [1983] method tries to find the best solution when the DM has an underlying pseudo-concave utility function. The method asks the DM tradeoff questions as well as questions involving choice between pairs of solutions. Adjacent efficient solutions in the presence and absence of previous responses are separately considered for preference questions. Questions about tradeoffs leading to these solutions are also asked. Using the responses of the DM the weights of a linear utility function are estimated which are then used to possibly jump to a preferred efficient solution. The method terminates at the global optimal solution if the DM does not like any of the solutions or tradeoffs presented. Termination occurs at a local optimal solution if the DM does not like any of the presented solutions but likes one or more of the presented tradeoffs.

Zionts and Wallenius [1983] solve a linear programming problem to find the weights of a linear utility function that is consistent with previous responses of the DM. The problem they solve finds the weights that are at maximum distance from the closest constraint. If the region defined by previous responses of the DM is empty then they delete the oldest response of the DM. The same procedure is used by Korhonen et.al. [1984] and Koksalan et.al. [1984] for discrete multiple criteria problems. Notice that if the last two responses of the DM are inconsistent with a linear utility function then all responses of the DM excluding the last response will be deleted.

In order to avoid the loss of information by deleting

previous responses we suggest a slightly different problem to solve in finding the weights:

$$
\begin{aligned}
&\text{Maximize } z^+ - z^- \\
&\text{subject to:} \\
&\lambda' C x^* - \lambda' C x^0 - z^+ + z^- \geq \epsilon \quad \text{for all } (x^*, x^0) \ni x^* > x^0 \\
&w_k \lambda - z^+ + z^- \geq \epsilon \quad \text{for all desired tradeoffs } w_k \\
&-w_i \lambda - z^+ + z^- \geq \epsilon \quad \text{for all undesired tradeoffs } w_i \\
&\lambda_j - z^+ \geq 0 \quad \forall j,\ j=1,2,\ldots,p \\
&\sum_{j=1}^{p} \lambda_j = 1 \\
&z^+ \geq 0,\ z^- \geq 0
\end{aligned}
$$

where C is the matrix of objective function coefficients and, x^* and x^0 are solutions in the decision variable space. λ is the vector of weights we are trying to find and ϵ is a small positive constant. z^+ and z^- stand for positive and negative deviations, respectively, from the constraints that are constructed using the responses of the DM.

Note that the above problem always has a feasible solution. The solution will force z^+ to be as large as possible and in case z^+ cannot be made positive z^- will be forced to be as small as possible. In other words, if there exists a feasible solution to the problem solved by Zionts and Wallenius [1983], above problem will find exactly the same solution. If such a solution does not exist then the above problem will find the solution that minimizes the maximum negative deviation from the responses of the DM. The idea we use somewhat resembles to goal programming.

We solved a number of problems having 10 to 30 constraints, 10 to 30 variables and 2 to 4 objectives, with the Z-W [1983] method and with its modified version. Both versions obtained the same solutions in all problems. Z-W [1983] had to delete old responses at most once in these problems. When old responses are deleted, Z-W [1983] usually found a poor solution as the solution that maximizes the utility function and had to ask the DM to compare that solution with the incumbent. The modified version, on the other hand, found the incumbent as the maximizing solution as well and avoided a comparison with a poorer solution. We expect that as the problem size grows Z-W [1983] method will have to delete more of the older responses and the savings in the number of comparisons using the modified version will increase.

3. A MODIFIED METHOD

The Z-W [1976] method requires the DM to answer tradeoff questions which are hard to answer for the DM. Furthermore, the method does not guarantee an improved solution in moving from an efficient solution to the next. The Z-W [1983] method tries to cope with a pseudo-concave utility function and this complicates the method both in terms of computations and the amount of preference information required. The method asks the DM his/her preference between pairs of distinct solutions as well as tradeoff questions. The method terminates at an extreme point solution which may be a local optimal solution.

We developed a simple approach by combining the above two

methods. Our method does not ask any tradeoff questions. The DM is required to compare only selected pairs of distinct efficient extreme point solutions in terms of their objective function values. The method moves to a more preferred solution at each iteration. If the DM has a linear utility function then the method finds the optimal solution. Otherwise the obtained solution may be local optimal.

The main idea of the method is to move to a preferred adjacent efficient solution until a solution that is preferred to all its adjacent efficient solutions is obtained. The steps of the method follow.

Step 1. Choose an arbitrary set of weights $\lambda' \geq 0$. Find the solution, x^*, for the problem

$$\text{Maximize } \lambda' C x$$
$$\text{s.t. } x \in X$$

where λ' is the weight vector, C is the matrix of objective function coefficients, x is the vector of decision variables and X is the feasible region.

Step 2. Identify solutions that are adjacent efficient to x^* in the presence of constraints from the previous responses of the DM.

Step 3. Ask the DM x^* vs an adjacent efficient solution. Do not ask the same question more than once. If the adjacent solution is preferred, make that solution the new x^* and go to 2. If x^* is preferred repeat this step. If x^* is preferred to all its adjacent solutions, stop.

In order to see how our approach compares with that of Zionts and Wallenius [1983] we solved 10 different problems that appeared in the literature. Number of constraints and variables range from 10 to 30, and the problems have 2, 3 or 4 objectives. We simulated the responses of the DM assuming an underlying utility function. The utility functions are:

(1) A weighted quadratic function

$$\underset{i}{\text{Minimize}} \sum_{j=1}^{p} w_j(z_j^i - z_j^*)^2$$

(2) A weighted Chebyshev function

$$\underset{i}{\text{Minimize}} \quad \{\underset{j}{\text{Max}} \{ w_j(z_j^* - z_j^i) \} \}$$

where p is the number of objectives, z_j^i is the value of objective function j at solution x^i, z_j^* is the maximum possible value of objective function j and w_j is the weight of objective function j. We solved each problem twice with each type of utility function using two different set of randomly generated weights. Therefore, with each of the two approaches we solved a total of 40 problems.

With both methods the global optimal solution is found in five of the 40 problems (the same five problems). In two of the remaining problems our method obtained better solutions and in three problems the Z-W [1983] method found better solutions. The amount of preference information required by our method was much less than that of the Z-W method. In most problems our method required less than half as many pairwise comparisons as the method of Z-W did.

The number and size of the problems we solved is not very representative to draw strong conclusions. However, our experience indicates that it may be possible to obtain as good solutions as does the method of Z-W [1983] in many problems with our simpler approach that requires much less information of the DM.

REFERENCES

[1] Green, P.E. and V. Srinivasan, "Conjoint Analysis in Consumer Research: Issues and Outlook", J. Consumer Research, Vol.5, 103-123, 1978.

[2] Koksalan, M.M., M.H. Karwan and S. Zionts, "An Improved Method for Solving Multiple Criteria Problems Involving Discrete Alternatives", IEEE Transactions on Systems, Man and Cybernetics, Vol.SMC-14, No.1, 24-34, 1984.

[3] Korhonen, P., J. Wallenius and S. Zionts, "Solving the Discrete Multiple Criteria Problem Using Convex Cones", Vol.30, No.11, 1336-1345, 1984.

[4] Miller, G.A., "The Magical Number Seven Plus or Minus Two: Some Limits on Our Capacity for Processing Information", Psychological Review, Vol.63, No.2, 81-97, 1956.

[5] Zionts, S. and J. Wallenius, "An Interactive Programming Method for Solving the Multiple Criteria Problem", Management Science,22, 652-663, 1976.

[6] Zionts, S. and J. Wallenius, "An Interactive Multiple Objective Linear Programming Method for a Class of Underlying Nonlinear Utility Functions", Management Science, 29, 519-529, 1983.

Prioritization Techniques Based on Stochastic Paired Comparisons

V. R. R. Uppuluri
Mathematical Sciences Section
Engineering Physics and Mathematics Division
Oak Ridge National Laboratory*
P.O. Box Y, Bldg. 9207A, MS 2
Oak Ridge, Tennessee 37831 U.S.A.

ABSTRACT

In 1977, Thomas L. Saaty introduced a method of assigning weights to categories based on data on paired comparisons. If we let a_{ij} denote the relative importance of the object being classified into the ith category, as opposed to the jth category, then it is assumed that $a_{ij} = 1/a_{ji}$. The normalized eigenvector associated with the largest eigenvalue of this reciprocal matrix is suggested as the vector of weights associated with the categories. In this paper, we extend this technique when the data is based on stochastic paired comparisons.

1. INTRODUCTION AND SUMMARY

In many unstructured decision problems it is of interest to know at least a rough ranking among the alternatives under consideration. For instance,

---

* Oak Ridge National Laboratory is operated by Martin Marietta Energy Systems, Inc. under Contract DE-AC-84OR21400 with the U.S. Department of Energy.

NATO ASI Series, Vol. F56
Multiple Criteria Decision Making and Risk Analysis
Using Microcomputers. Edited by B. Karpak and S. Zionts

in national planning, one may be interested in setting priorities among a class of energy options. In industry, a manufacturer may have to decide which products he must produce based on the opinions of some judges who were given some samples. At times, when there are several products of the same item available, one may like to know the best from the worst based on the opinions of some experts. In this report we shall consider analysis of data, based on paired comparisons, in the context of some of these problems.

In 1977, T. L. Saaty introduced a methodology which can be used to rank k objects based on paired comparisons; and he found several applications discussed in two later books (1980, 1982). The method of Saaty requires that we compare the objects two at a time and determine how important one object is relative to the other. In this report, we modify this procedure to the case of paired comparisons where one needs to determine only whether one object is more or less important or equally important as the other object. As a further modification, we also consider the case of paired comparisons where we have data on the probability that one object is more important than another object. These modification are discussed in sections 3 and 4 with examples. We also consider the case when we have data from several judges and illustrate the method in section 3 with examples.

2. TWO PROCEDURES FOR RANKING OBJECTS

Suppose we have k objects and we wish to rank them according to a certain characteristic. We assume that a judge compares these objects two at a time, and says that the object i is $0 < a_{ij}$ times as important as object j; and this in the mind of the judge is equivalent to saying that object j is $1/a_{ij}$ times as important as object i. Thus, based on paired comparisons of the k objects, a judge provides the data a_{ij} which can be represented by the $k \times k$ reciprocal matrix ($a_{ij} = 1/a_{ji}$, $1 \leqslant i,\ j \leqslant k$). Given this data, there are two procedures which will enable us to rank the objects.

Procedure 1. λ_{max} Procedure

In 1977, Thomas L. Saaty (1977) suggested a method of assigning normalized weights to k objects, bassed on the eigenvector associated with the largest eigenvalue of the reciprocal paired comparison matrix. By Perron-Froebenius theorem, we have the existence of the largest eigenvalue (λ_{max}) of a reciprocal matrix, and know it to be real and positive; we also know that the components of the eigenvector $(x_1, x_2, \cdots, x_k)'$ associated with this λ_{max} to be real and positive. Saaty suggested to assign the weight $w_i = x_i/(x_1 + \cdots + x_k)$ to the ith object, and then one can rank the objects according to the ranks of $w_1, w_2, \cdots, w_k$.

Procedure 2. Log Least Squares Procedure

We shall now describe a procedure of assigning weights on k objects based on the log linear model

$$ln\ a_{ij} = ln\ \omega_i - ln\ \omega_j + \epsilon_{ij}\ .$$

The data based on the paired comparisons is represented by the $k \times k$ reciprocal matrix

$$\begin{bmatrix} a_{11} & a_{12} & a_{13} & \cdot & \cdot & \cdot & a_{1k} \\ a_{21} & a_{22} & a_{23} & \cdot & \cdot & \cdot & a_{2k} \\ \cdot & \cdot & \cdot & \cdot & \cdot & \cdot & \\ \cdot & \cdot & \cdot & \cdot & \cdot & \cdot & \\ \cdot & \cdot & \cdot & \cdot & \cdot & \cdot & \\ a_{k1} & a_{k2} & a_{k3} & \cdot & \cdot & \cdot & a_{kk} \end{bmatrix}$$

where $a_{ii} = 1$ and $a_{ij} = 1/a_{ji}$. Let

$$r_1 = (a_{11} \quad a_{12} \quad \cdots \quad a_{1k})^{1/k}$$
$$r_2 = (a_{21} \quad a_{22} \quad \cdots \quad a_{2k})^{1/k}$$
$$r_i = (a_{i1} \quad a_{i2} \quad \cdots \quad a_{ik})^{1/k} \qquad \text{for } 1 \leqslant i \leqslant k.$$

The logarithmic least squares estimates of ω_i/ω_j are given by

$$\omega_i/\omega_j = r_i/r_j \ ,$$

and the normalized weight assigned to the ith object is given by

$$\omega_i = r_i/(r_1 + r_2 + \cdots + r_k) \quad \text{for } i = 1,\ 2,\ \cdots,\ k\ .$$

The k objects are ranked according to the ranks of $\omega_1, \omega_2, \cdots, \omega_k$.

Several applications on the $\lambda_{\max}$ method appear in numerous publications by Saaty and coworkers, and some of them were presented in the two books by Saaty (1980, 1982). At Oak Ridge, an application of this method was made by Holland and Uppuluri (1979).

3. SIMPLE COMPARISONS BETWEEN OBJECTS

When there are several objects (k large) to be ranked it is difficult to obtain data on a_{ij} which indicates the importance of object i compared with object j. In this section, we consider the situation when object i is better than object j in which case we assign $a_{ij} = e$, (and $a_{ji} = 1/e$) or object i is as good as object j in which case we assign $a_{ij} = 1$. For such data we use the procedures discussed in the previous section to assign ranks to the k objects.

Example 3.1 Arrow's Paradox.

Suppose we have three objects A, B, and C, and a judge finds: A is better than B, B is better than C, and C is better than A, then we have the following reciprocal matrix

$$
\begin{array}{c} \\ A \\ B \\ C \end{array}
\begin{array}{c}
\begin{array}{ccc} A & B & C \end{array} \\
\left[\begin{array}{ccc}
1 & e & 1/e \\
1/e & 1 & e \\
e & 1/e & 1
\end{array}\right]
\end{array}
$$

Since $1 + e + 1e$ is the largest eigenvalue λ_{max} of this matrix and [1.3, 1/3, 1/3] is the normalized eigenvector associated with λ_{max}, we conclude that the three objects A, B, and C are equally good (or equally bad).

<u>Example 3.2 Chess Players</u>.

We shall consider the problem of six players A, B, C, D, E, and F engaged in a chess tournament in which each plays the other once. The following data was discussed by T. H. Wei in a Ph.D. Thesis at the University of Cambridge, England. Player A beats Player C (which is denoted by e) and Player A loses to Player D (denoted by 1/e), etc., is represented in the matrix form:

$$
\begin{array}{c} \text{Player} \\ A \\ B \\ C \\ D \\ E \\ F \end{array}
\begin{array}{c}
\begin{array}{cccccc} A & B & C & D & E & F \end{array} \\
\left[\begin{array}{cccccc}
1 & e & e & 1/e & e & e \\
 & 1 & 1/e & e & e & 1/e \\
 & & 1 & e & e & e \\
\text{"Reciprocal"} & & & 1 & 1/e & 1/e \\
 & & & & 1 & e \\
 & & & & & 1
\end{array}\right]
\end{array}
$$

For this problem, the λ_{max} method gives the normalized weights

$$w_A = .2486, \quad w_B = .1341, \quad w_C = .2197,$$
$$w_D = .1294, \quad w_E = .1341, \quad w_F = .1341,$$

and we conclude that $A > C > B \approx E \approx F > D$. This ranking agrees with the ranking given by Wei, and discussed by Kendall (1955).

We shall now consider the case when we have several judges. Let $a_{ij}(\alpha)$ denote the paired comparison of object i with object j, by judge α. If we have m judges we obtain the "average matrix" with elements $a_{ij}(*)$ defined by

$$a_{ij}(*) = \left\{ \prod_{\alpha=1}^{m} a_{ij}(\alpha) \right\}^{1/m}$$

In order to obtain a ranking of the k objects, based on the data of the m judges we use the $\lambda_{\max}$ procedure on the "average matrix." We shall illustrate this method in Examples 3.3, 3.4, and 3.5.

Example 3.3.: 16 Samples, 3 Judges.

A manufacturer has 16 samples (candidates) and wants to know a ranking from best to worst. He asks three judges to compare the 16 samples two at a time. Judge I, choses sample 1 over all the others, and chooses sample 2 over all the others, etc., and Judge II choses sample 2 over sample 1, sample 1 over sample 3, etc., and Judge III chooses sample 1 over samples 2 to 15, and sample 16 over sample 1, etc. This data is presented in Table 3.3.1.

Based on the geometric mean method suggested above, we obtain the average matrix given in Table 3.3.2.

For this "average matrix," the $\lambda_{\max}$ method assigns the following ranks (from best to worst): sample 3 > sample 9 > sample 10 > $\cdots$ > sample 1 > sample 16. This ranking shown in Table 3.3.3 is in agreement with the ranking that was used by the manufacturer based on the traditional procedures used at their plant.

JUDGE I

Candidate	1	2	3	4	5	6	7	8	9	10	11	12	13	14	15	16
1	1	e	e	e	e	e	e	e	e	e	e	e	e	e	e	e
2		1	e	e	e	e	e	e	e	e	e	e	e	e	e	e
3			1	1/e	1/e	1/e	1/e	1/e	1/e	1/e	1/e	1/e	1/e	1/e	1/e	1/e
4				1	1/e	e	e	1/e	e	e	e	e	1/e	e	1/e	1/e
5					1	e	e	e	e	e	e	e	1/e	e	e	1/e
6						1	e	1/e	e	e	e	e	1/e	e	1/e	1/e
7							1	1/e	e	1/e	1/e	e	1/e	e	1/e	1/e
8								1	e	e	e	e	1/e	e	1/e	1/e
9									1	e	1/e	e	1/e	1/e	1/e	1/e
10										1	1/e	e	1/e	e	1/e	1/e
11											1	e	1/e	e	1/e	1/e
12												1	1/e	1/e	1/e	1/e
13													1	e	e	e
14														1	1/e	1/e
15															1	1/e
16																1

JUDGE II

Candidate	1	2	3	4	5	6	7	8	9	10	11	12	13	14	15	16
1	1	/e	e	e	e	e	e	e	e	e	e	e	1/e	e	1/e	1/e
2		1	e	e	e	e	e	e	e	e	e	e	1/e	e	e	1/e
3			1	1/e	1/e	1/e	1/e	1/e	e	1/e	1/e	1/e	1/e	e	1/e	1/e
4				1	1/e	e	e	e	e	e	e	e	1/e	e	1/e	1/e
5					1	e	e	1/e	e	e	e	e	1/e	e	e	1/e
6						1	e	1/e	e	e	e	e	1/e	e	1/e	1/e
7							1	1/e	e	e	1/e	1/e	1/e	e	1/e	1/e
8								1	e	e	e	e	1/e	e	1/e	1/e
9									1	e	1/e	1/e	1/e	1/e	1/e	1/e
10										1	1/e	e	1/e	e	1/e	1/e
11											1	e	1/e	e	1/e	1/e
12												1	1/e	e	1/e	1/e
13													1	e	e	1/e
14														1	1/e	1/e
15															1	1/e
16																1

JUDGE III

Candidate	1	2	3	4	5	6	7	8	9	10	11	12	13	14	15	16
1	1	e	e	e	e	e	e	e	e	e	e	e	e	e	e	1/e
2		1	e	e	e	e	e	e	e	e	e	e	e	e	e	1/e
3			1	1/e	1/e	1/e	1/e	1/e	1/e	1/e	1/e	1/e	1/e	1/e	1/e	1/e
4				1	1/e	e	e	1/e	e	e	e	e	1/e	e	1/e	1/e
5					1	e	e	e	e	e	e	e	1/e	e	1	1/e
6						1	e	1/e	e	e	e	e	1/e	e	1/e	1/e
7							1	1/e	e	e	1/e	1/e	1/e	1	1/e	1/e
8								1	e	e	e	e	1/e	e	1/e	1/e
9									1	e	1/e	1/e	1/e	1/e	1/e	1/e
10										1	1/e	1/e	1/e	1/e	1/e	1/e
11											1	e	1/e	e	1/e	1/e
12												1	1/e	e	1/e	1/e
13													1	e	e	1/e
14														1	1/e	1/e
15															1	1/e
16																1

3.3.1 Data from 3 Judges

Candidate	1	2	3	4	5	6	7	8	9	10	11	12	13	14	15	16
1	1.0000	0.7165	0.3679	0.3679	0.3679	0.3679	0.3679	0.3679	0.3679	0.3679	0.3679	0.3679	0.7165	0.3679	0.7165	1.3966
2		1.0000	0.3679	0.3679	0.3679	0.3679	0.3679	0.3679	0.3679	0.3679	0.3679	0.3679	0.7165	0.3679	0.3679	1.3966
3			1.0000	2.7183	2.7183	2.7183	2.7183	2.7183	1.3956	2.7183	2.7183	2.7183	2.7183	1.3956	2.7183	2.7183
4				1.0000	2.7183	0.3679	0.3679	1.3956	0.3679	0.3679	0.3679	0.3679	2.7183	0.3679	2.7183	2.7183
5					1.0000	0.3679	0.3679	0.7165	0.3679	0.3679	0.3679	0.3679	2.7183	0.3679	0.5134	2.7183
6						1.0000	0.3679	2.7183	0.3679	0.3679	0.3679	0.3679	2.7183	0.3679	2.7183	2.7183
7							1.0000	2.7183	0.3679	0.7165	2.7183	1.3956	2.7183	0.5134	2.7183	2.7183
8								1.0000	0.3679	0.3679	0.3679	0.3679	2.7183	0.3679	2.7183	2.7183
9									1.0000	0.3679	2.7183	1.3956	2.7183	2.7183	2.7183	2.7183
10										1.0000	2.7183	0.7165	2.7183	0.7165	2.7183	2.7183
11											1.0000	0.3679	2.7183	0.3679	2.7183	2.7183
12												1.0000	2.7183	0.7165	2.7183	2.7183
13													1.0000	0.3679	0.3679	1.3956
14														1.0000	2.7183	2.7183
15															1.0000	2.7183
16																1.0000

3.3.2 Average Matrix

1.39561	2.71828	0.36788	0.36788	1.94773	0.36788	0.36788
0.36788	0.36788	0.36788	0.36788	0.36788	2.71828	0.36788
1.00000	2.71829					

0.71653	0.71653	0.36788	0.36788	0.36788	0.36788	0.36788
0.36788	0.36788	0.36788	0.36788	0.36788	0.71653	0.36788
0.36788	1.00000					

THE FINAL WEIGHTS ARE:

0.02500	0.02523	0.12685	0.04857	0.03685	0.05876	0.08452
0.04402	0.10961	0.10045	0.06693	0.08836	0.02622	0.09821
0.03711	0.02333					

Total: 1.00000

Candidate:	1	2	3	4	5	6	7	8
Weight:	0.0250	0.0252	0.1268	0.0486	0.0368	0.0588	0.0845	0.0440
Rank:	15	14	1	9	12	8	6	10

Candidate:	9	10	11	12	13	14	15	16
Weight:	0.1096	0.1004	0.0669	0.0884	0.0262	0.0982	0.0371	0.0233
Rank:	2	3	7	5	13	4	11	16

3.3.3. Final Ranking

4. STOCHASTIC COMPARISONS BETWEEN OBJECTS

In this section, we consider the situation when we have stochastic information on paired comparisons, and we still wish to rank the k objects. Suppose we have a dice A with six faces, marked on 0 with two faces and marked with 4 on four faces, i.e., $A = (0, 0, 4, 4, 4, 4)$, and let dice B be marked with 3 on all its six faces, i.e., $B = (3, 3, 3, 3, 3, 3)$. We say that dice A beats dice B, whenever we see a higher number on dice A on one roll of both the dice. On some rolls, we get a higher number on dice B, but the probability that A beats B is 2/3. Suppose we have two more dice C and D where $C = (2, 2, 2, 2, 6, 6)$ and $D = (1, 1, 1, 5, 5, 5)$. We can easily compute the several probabilities with which the dice beat each other; these are given in Matrix 4.1.

Dice	A	B	C	D
A	1/2	2/3	4/9	1/3
B	1/3	1/2	2/3	1/2
C	5/9	1/3	1/2	2/3
D	2/3	1/2	1/3	1/2

Matrix 4.1 Probability matrix of 4 Dice

Now, given the stochastic comparison matrix of the four dice A, B, C, and D, the problem is to rank the dice. We suggest the following method to obtain a 4×4 reciprocal matrix.

If the probability that the ith dice beats the jth dice is equal to θ, let $a_{ij} = \exp(2\theta-1)$ and $a_{ji} = 1/a_{ij} = \exp(1-2\theta)$. Thus from matrix (4.1), we obtain the reciprocal matix (4.2) given below:

	A	B	C	D
A	1	$e^{1/3}$	$e^{-1/9}$	$e^{-1/3}$
B	$e^{-1/3}$	1	$e^{1/3}$	1
C	$e^{1/9}$	$e^{-1/3}$	1	$e^{1/3}$
D	$e^{1/3}$	1	$e^{-1/3}$	1

Reciprocal Matrix 4.2.

We apply the λ_{max} procedure now and find $\lambda_{max} = 4.11$ and tl normalized eigenvector to be (.2435 for A, .2511 for B, .2567 for C, .24ε for D), and the final ranking to be $C > B > D > A$.

It is clear that this procedure could also be incorporated if we h‵‿ several judges, or if we have a hierarchical sheme of stochastic comparisons.

5. REMARKS

Several computer programs were developed at Oak Ridge which can be used to rank up to 50 objects based on paired comparisons data. The λ_{max} algorithm was developed at Oak Ridge, which efficiently computers the largest eigenvalue fo a reciprocal matrix without finding the whole spectrum of the matrix. This procedure utilizes the log least squares estimated values as the initial eigenvector, which makes the computer algorithms to find λ_{max} accurate and efficient.

Some of the computer programs are of conversational mode and investigators can sit in front of the terminal and obtain ranks of objects.

Further research needs to be done in understanding the uncertainties associated with the weights associated with the objects based on the procedures discussed above.

6. ACKNOWLEDGEMENTS

We wish to thank Mr. Elmon Leach, who developed the computer programs used in this report. Thanks are due to Dr. David S. Scott, for developing an efficient algorithm to find the largest eigenvalue of a reciprocal matrix.

REFERENCES

[1] Holland, C. W. and Uppuluri, V. R. R. (1979), "Application of Saaty's Method to Managerial Selection," Y/EX-9, Union Carbide Nuclear Division Report, Oak Ridge, Tennessee.

[2] Kendall, M. G. (1955), "Further Contributions to the Theory of Paired Comparisons," *Biometrics*, **11**, 43-62.

[3] Saaty, Thomas L. (1977), "A scaling Method for Priorities in Hierarchical Structures," *J. Math. Psychology*, **15**, 234-281.

[4] Saaty, Thomas L. (1980), *The Analytic Hierarchy Process*, McGraw-Hill, New York, New York.

[5] Saaty, Thomas L. (1982), *Decision Making for Leaders*, Lifetime Learning Publications, Belmont, California.

ON CAPACITY PLANNING OF DIAGNOSIS SYSTEMS IN HOSPITALS

Günter Fandel and Holger Hegemann*
Chair for Business Administration
Fernuniversität
Feithstr.140, D-5800 Hagen 1, FRG

1. Object and course of the investigation

In the last twenty years hospitals have developed into cost and investment intensive service enterprises confronted with a rising pressure of costs. Apart from an enormous expansion in personnel costs the reasons can be seen in a higher equipment with apparatus based on the rapid developments in the medical-technical progress and the therewith attendant and fast increasing demand for hospital services.

Aspects of medical necessity and effectiveness are decisive for the production of health services in hospitals; but this process must be reviewed on the basis of efficiency considerations as well. This follows from the mere fact that a hospital must be regarded as a consumer oriented facility having to provide certain medical services efficiently by combining different production factors. Although some methodical and legal deliberations have already been made in this direction, the German hospital system does not yet satisfy those economic criteria being already fulfilled in other industrialized nations.

---

* The authors are especially indebted to the former medical director, Dr. Klaus Pieper, and the team of doctors of the diagnosis system at the General Hospital of Hagen City for having made possible the empirical study carried out in this context.

NATO ASI Series, Vol. F56
Multiple Criteria Decision Making and Risk Analysis
Using Microcomputers. Edited by B. Karpak and S. Zionts

In this context the fact is characteristic, that hospital management was not represented at the health economy congress of the Verein für Socialpolitik 1985 by a particular study group.

This argumentation applies correspondingly to the diagnostics being of central significance to the service production in hospitals. For, this subsystem is claimed as the first service area by the patients and the examination results received hereby give important orientation marks for the following therapy and care.

The inexpedient allocation of medical staff and equipment especially in the diagnostics would lead to considerable differences in the load degrees of the different diagnostic service units and to increased waiting times for the patients in front of the diagnostic rooms. The latter would entail a retardation of the following therapy and care. In order to avoid such misallocations and their consequences capacity planning in the clinical diagnostics on the basis of the methods of operations research is indispensable.

So far, only subsystems of the clinical diagnostics have been analyzed in the literature by means of network-planning-technique (Fleck 1977, Taylor/Keown 1980), queuing theory (Bailey 1954, Taylor/Templeton 1980) and simulation (Carruthers 1970, Gierl 1976, O'Kane 1981, Revesz/Shea/Ziskin 1972). The above mentioned problem, however, can only be solved by a central capacity planning including the interdependencies existing in the form of patients' ways among the single diagnostic units of a hospital. In addition, under instrumental aspects it must be objected, that the network-planning-technique does not allow the consideration of the patient's arrival behavior in the diagnostic area and that the simulation as a method without the quality of convergency will remain inferior to the queuing theory as long as the special planning situation allows the application of this exact method.

Consequently, a queuing model, which shall describe the structure of the diagnostic system and the diagnosis process by means of patients flows as close to reality as possible, is chosen as the basis for the analytical investigations to be carried out within the scope of capacity planning. In contrast to the approaches in the literature, in which only single diagnostic rooms or units and actually only one patient flow were objects of the model considerations, here however two modifications are necessary. On the one hand, the entire approach demands that a network-oriented total system of the diagnostics is taken as the basis. On the other hand, attention has to be paid to the fact that this general system is passed through by two patient streams simultaneously resulting from the examination processes of out-patients and in-patients. With regard to the management and optimal configuration of such systems it can be demonstrated, though, that the queuing models well-known from the literature (Gross/Harris 1974, Jackson 1963) can also be extended easily to such a generalized problem situation under certain assumptions (Hegemann 1985).

The queuing model for capacity planning presented in the following can be applied to the planning of measures for constructing new or extending old buildings in the clinical diagnostics of hospitals. At the same time, it allows to reveal weaknesses within existing diagnosis systems by means of system indices like patients' waiting times, load degrees of the apparatus and the length of the queues and to show alternatives for an improvement of the capacity structure of the diagnosis area. Thereby the long-term optimal capacity structure is often characterized by waiting times of patients in front of diagnostic service units not exceeding a given reasonable limit and reaching a possibly high load degree for the single examination rooms as well as for the total diagnostic system.

Finally, the methodical reflections will be exemplified for the diagnosis system of the General Hospital of Hagen City, which contains 650 beds. Thoughts on rationalization are in the foreground of the application-oriented statements. It will appear, that simple organizing measures concerning the rearrangement and fusion of diagnostic service units can reduce the average patients' waiting times by more than 20% and allow an increase in the expected load degree of the total diagnosis system by 18%.

2. Queuing theoretical description of a diagnosis system

2.1 An outline of the diagnostics and their particularities

Beside therapy, care, and medical education and research the diagnostics are part of the medical area of a hospital, which is differentiated from the economic and administrative area containing mainly management tasks. The numerous relations possibly existing between the diagnosis system and other subareas of a hospital are not object of the reflections. The considerations rather concentrate on the diagnostics as a system, their system elements and the interdependencies occurring among them as well as on the derivation of suitable system indices characterizing the quality of the diagnosis system under economic aspects. The diagnostics possess their external relations in the flows of out-patients and in-patients entering, passing through and leaving the diagnosis system for reasons of diagnosis on the initiative of private doctor's practices or by order of the care and therapeutical departments of the hospital.

The diagnostics can be viewed as the totality of all those activities necessary for the production of a diagnosis; so these activities delineate the way to a diagnosis. Under system analytical aspects the diagnostics are usually divided into the sections X-ray diagnostics, laboratory diagnostics, function diagnostics, and endoscopy (Eichhorn 1976, p. 281), which themselves can be subdivided into single service units or diagnosis rooms. In these service units being elements of the diagnosis system single diagnostic services are produced in the form of examinations by combining medical personnel, diagnostic equipment and means of medical needs as resources. The medical staff consists of doctors, medical-technical assistants, medical helpers as well as, occasionally, laboratory assistants and operating nurses. The actual diagnosis generally emerges from the composition of such individual diagnostic services in different service units; so the diagnosis ensues within the scope of a diagnostic process or is its results, respectively.

Within the diagnosis system the service units, generally united in departments under the disciplinary responsibility of a head physician, are connected by the flows of in-patients and out-patients and the information flux accompanying them. These flows result from the diagnostic processes arranged by doctors of the different medical departments in dependence of the sickness of the individual patient to set up the diagnosis. These processes, normally different for every patient, can be seen as disease-specific parts of the total diagnostic process because of their disjunctive connection of individual diagnostic examinations. Usually they are not fixed a priori, but are mostly made concrete by the doctors during the examination. So the patients' ways via the diagnostic service units, essentially determining the internal relation structure of the total diagnosis system, must be considered as stochastic.

It is obvious to understand the diagnosis system as a waiting system and therefore to make it accessible to a queuing theoretical analysis. The diagnostic rooms can be interpreted as service stations, the patients are the traffic units, the waitingroom is the queue, and the flows of the out-patients and in-patients correspond to the traffic streams. Thus it is possible to describe the formal structure of the diagnosis system as a discrete, dynamic, stochastic, and open system by the number of diagnostic service units, the stochastic arrival behavior of the out-patients and in-patients, the stochastic examination times in the service units as well as by the stochastic patients' ways with the transition probabilities from one service unit to the next. This, however, is not possible in the scope of a conventional queuing model; on the contrary, two extensions are necessary. Firstly, it has to be taken into account that the demand flow consists of two partial streams. Secondly, the stochastic structure of the diagnosis processes being dependent on the patients' ways has to be regarded. Whereas directly usable clues can be found in the queuing theoretical literature for the first modification, which can be taken over as results here, the necessary second modification is mastered by adding the GERT network-model (Pritsker/Happ 1966, Neuman/Steinhardt 1979) to the queuing model in order to conceive realistically the stochastic arrival processes in the form of patients' ways. To a certain extent, both planning procedures are combined by including the exclusive-or-branchings known from the GERT model and therewith the transition probabilities of the patients between the diagnosis rooms into the queuing theoretical model construction. On this basis realistic system indices can be determined for the total diagnosis system giving information about how well the single service units are coordinated and allowing secure statements about the service quality and the efficiency of the total diagnosis system and its partial elements.

In order to judge about the efficiency and service quality of a queuing system diagnostics the following system indices for the single service units i, i=1, . . ., N, as well as for the total system are of interest:

(1) the expected waiting time of a patient in front of a diagnostic room and in the total system; it may be denoted by $E(V_{q_i})$ and $E(V_q)$.

(2) the expected time a patient spends in a service unit and in the total system; it is indicated by $E(V_i)$ and $E(V)$.

(3) the expected number of waiting patients in a service unit and in the total system; it is expressed by $E(L_{q_i})$ and $E(L_q)$.

(4) the expected number of patients in a service unit and in the total system; it is symbolized by $E(L_i)$ and $E(L)$.

(5) the expected load degree of a service unit and of the total system; it will be denoted by ξ_i and ξ.

These indices serve also as a basis for changes in the capacity structure of a diagnosis system to be carried out eventually.

Among these system indices the expected waiting time of a patient and the expected load degree of the service units and the total system are relevant with regard to health policy. These two criteria, which are to be minimized or maximized, are in conflict with each other being known for analogous areas of industrial production as the dilemma of scheduling. In the field of health policy this conflict is mostly solved in the way that the load degrees are maximized under the constraint that the waiting times do not exceed socially reasonable limits.

2.2 Formal description of the model of a diagnosis system

2.2.1 Premises

The queuing theoretical analysis to be carried out later on is based on the following premises:

(P 1) The out-patients and in-patients arrive at the diagnosis system from outside one by one; the arrival processes are Poisson processes with the intensities λ_0^a (for the out-patients) and λ_0^s (for the in-patients), i.e. the interarrival times are exponentially distributed with the parameters λ_0^a and λ_0^s. The subindex zero indicates the outside of the diagnosis system as the source and sink of the patients flows.

(P 2) The number of potential patients, who can enter the diagnosis system, is regarded as unlimited.

(P 3) The patients remain in the queue of the respective diagnostic service unit they joined after their arrival up to the beginning of the examination.

(P 4) The examinations of the patients in the respective diagnosis rooms take place individually and are not dependent on the arrival processes.

(P 5) The examination times of the diagnostic actions in the service units are exponentially distributed with the parameters μ_i, $i=1, . . . ,N$.

(P 6) An examination once started will be finished without interruptions.

(P 7) Breakdown periods of the equipment are not considered.

(P 8) The diagnosis system consists of N service units; each diagnostic service unit i, i=1, . . . ,N, can be composed of $n_{p_i} \in \mathbb{N}$ parallel and identical diagnostic rooms.

(P 9) If a patient has once left a diagnostic service unit i, he cannot enter it immediately again, i.e. the transition probability r_{ii} is equal to zero for all $i \in \{1, . . ., N\}$.

(P 10) The ways of patients from a service unit i to a service unit j are independent of each other for all i,

$j \in \{1, . . . ,N\}$, $i \neq j$.

(P 11) Times needed for the ways between service units may be neglected.

(P 12) The way of a patient through the diagnosis system is free of cycles; so he enters each service unit once at most.

(P 13) Waiting-room capacities in the diagnostic service units are unlimited.

(P 14) Patients enter the diagnosis rooms according to the service discipline "first come - first served" (FCFS).

(P 15) There are two independent sources of the flows of out-patients and in-patients.

The premises (P 1) and (P 5) were backed by the results of the empirical study - that will be reported later -, so that their supposition which can also be found in the literature is of no problem. In addition (P 5) allows the arbitrary joining or separation of diagnostic examinations without changing their distribution character. This is relevant for an eventual organizational rearrangement of diagnostic examinations necessary to reach an optimal capacity structure; that will be discussed more detailed below.

(P 9) and (P 12) exclude backward loops which would contradict (P 4). Otherwise the arrival processes in front of the service units would no longer be Poisson processes (Walrand 1982, Malamed 1979a, 1979b); this quality, however, is decisively necessary for the following analysis. Moreover, (P 12) was backed by the empirical study.

2.2.2 The formal model of a general diagnosis system and its characteristics

Under the premises stated above the diagnosis system can be understood as a network-type queuing system of $N \in \mathbf{N}$ service units i, i=1, . . . ,N, arranged in parallel or in series, which is run through by q=2 different patient flows as graphically demonstrated in Figure 1. In this connection

- λ_i^a, λ_i^s and λ_j^a, λ_j^s, respectively, indicate the flows of out-patients and in-patients going to and leaving the outside of the diagnosis system and the diagnostic service units in the diagnosis system with i, j $\in$ {0, . . . ,N}.

- r^a_{ij}, r^s_{ij} and r^a_{ji}, r^s_{ji}, respectively, indicate the transition probabilities of the out-patients and in-patients from unit i to unit j and vice versa with i, j $\in$ {0, . . . ,N}, $r^a_{ij}= r^a_{ji}= r^s_{ij}= r^s_{ji}= 0$ for i=j (premise (P 9)) and

$$\sum_{j=0}^{N} r^a_{ij} = \sum_{j=0}^{N} r^s_{ij} = 1, \quad i=0, . . . ,N.$$

On account of the premises (P 1), (P 5), (P 8), (P 13) to (P 15) the diagnosis system can be typified extending Kendall's notation (1953) by the queuing model M/M/N/∞/FCFS/2; the first two symbols indicate that the distributions of the patients' interarrival and examination times are Markovian. If there is only one diagnostic service unit with one diagnosis room or n_p parallel diagnosis rooms, one gets the well-known model typifications M/M/1/∞/FCFS/2 and M/M/n_p/∞/FCFS/2,respectively.

Furthermore the general diagnosis system is characterized by the fact, that a branching of the patients' flows in the meaning of an exclusive-or-node appears behind the outside of the diagnosis system and every diagnosis room and that a fusion or a superimposition of several patients' flows occurs in front of every diagnostic service unit and the outside of the diagnosis system. According to the patient's type these flows are ordered in out-patients and in-patients. Because of the following three basic theorems all (partial) flows of out-patients and in-patients within the diagnostic system are Poisson processes.

Theorem 1 (Burke 1956, 1972):

The departure processes of an M/M/1/∞/FCFS/1- and an M/M/n_p/∞/FCFS/1-queuing system being in equilibrium are Poisson streams and have the same intensities as the arrival streams.

Figure 1: The diagnosis system as a queueing system

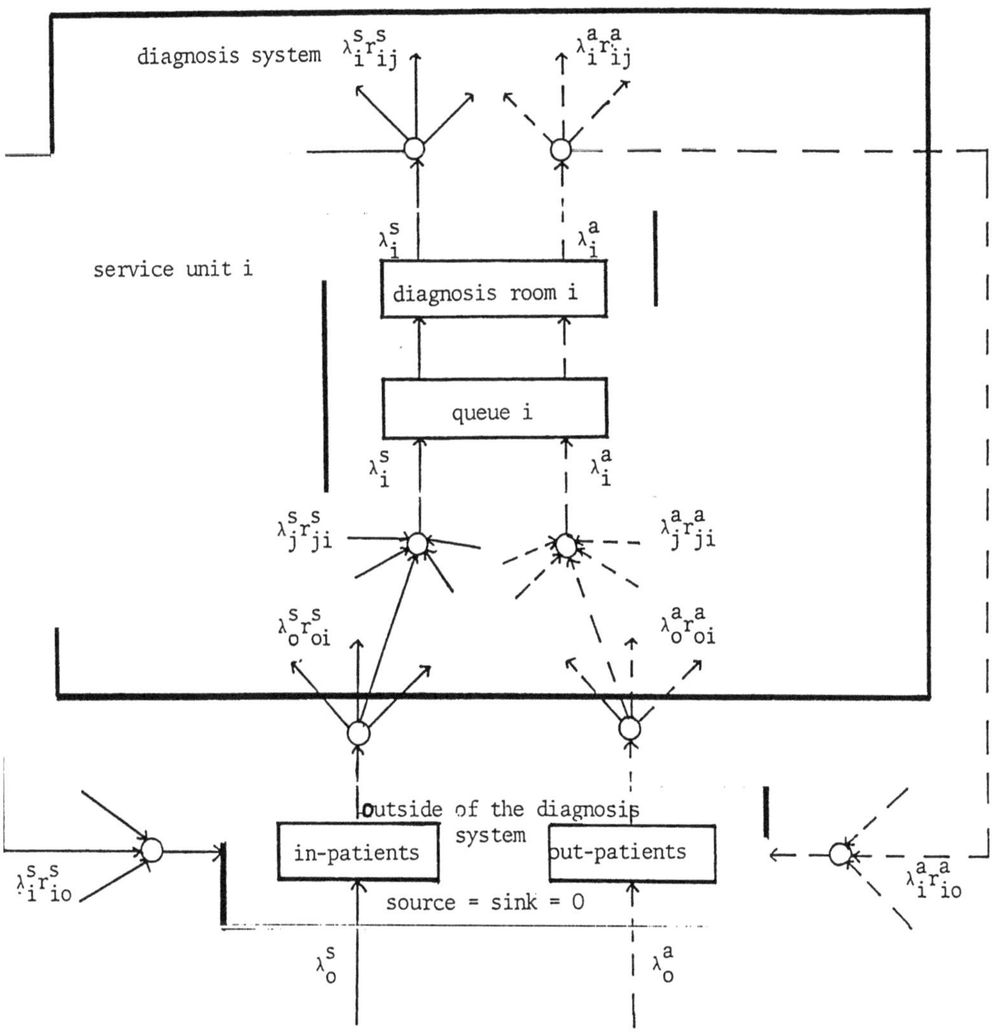

Theorem 2 (Klimow 1979):

The composition of c independent Poisson streams with the intensities $\lambda_1, \ldots, \lambda_c$, $c \in \mathbf{N}$, is again a Poisson stream with the intensity $\lambda = \lambda_1 + \ldots + \lambda_c$.

Theorem 3 (Ferschl 1964):

If the mobile units of a Poisson stream with the intensity λ are stochastically distributed over $c \in \mathbf{N}$ partial streams in such a way that subsequently a unit is within the partial stream i with the probability r_i $(i=1, \ldots, c)$, then the c partial streams are Poisson streams each with the intensity $r_i\lambda$.

This fact is of advantage for the following parts of the analysis. On the one hand it is possible to determine the arrival rates λ_i in front of the single diagnostic service units just by means of the transition probabilities r^a_{ij} and r^s_{ij} together with the total arrival rates λ^a_0 and λ^s_0. On the other hand this firstly allows to analyze the single diagnostic service units separately by applying queuing theoretical considerations and to describe their efficiency by means of indices before, in a second step, one obtains the indices of the total diagnosis system from the combination of the indices of the single diagnostic service units.

Now, $\lambda^a = (\lambda^a_0, \ldots, \lambda^a_N)^T$ may denote the vector of the arrival rates of out-patients and $\lambda^s = (\lambda^s_0, \ldots, \lambda^s_N)^T$ that of in-patients, and

$$R^a = \begin{pmatrix} 0 & r^a_{01} & \cdots & r^a_{0N} \\ r^a_{10} & 0 & \cdots & r^a_{1N} \\ \vdots & & & \vdots \\ r^a_{N0} & & \cdots & 0 \end{pmatrix} \text{ and } R^s = \begin{pmatrix} 0 & r^s_{01} & \cdots & r^s_{0N} \\ r^s_{10} & 0 & \cdots & r^s_{1N} \\ \vdots & & & \vdots \\ r^a_{N0} & & \cdots & 0 \end{pmatrix}$$

may indicate the so-called routing matrices of the out-patients and the in-patients with the corresponding transition probabilities. Then the relations

$$\begin{aligned} \lambda^a &= R^{a^T}\lambda^a, \\ \lambda^s &= R^{s^T}\lambda^s, \\ \lambda &= \lambda^a + \lambda^s = R^{a^T}\lambda^a + R^{s^T}\lambda^s \end{aligned}$$

are obviously valid. If λ_0^a and λ_0^s are known it is easily possible to calculate the flow intensities λ_i, λ_i^a and λ_i^s needed for the analysis from these equation systems by means of the transposed routing matrices of the transition probabilities. In case the single diagnostic service units and the total diagnostic system are in equilibrium, i.e.

$\lambda_i^a + \lambda_i^s < \mu_i$ and $\lambda_i^a + \lambda_i^s < n_{p_i}\mu_i$, respectively,

holds for every service unit with only one or n_{p_i} parallel diagnosis rooms, one obtains from the equations the queuing theoretical indices concerning the single diagnostic service units and the total system as mentioned in Paragraph 2.1 and compiled in Table 1. A single service unit corresponds - as already explained - to an M/M/1/∞/FCFS/2- or an M/M/n_p/∞/FCFS/2-model. Furthermore, the total system M/M/N/∞/FCFS/2 is also to be differentiated whether the number of parallel, identical

Table 1: Indices of the single service units and the total diagnosis system

indices	single service unit M/M/1/∞/FCFS/2	M/M/n_p/∞/FCFS/2	total system M/M/N/∞/FCFS/2 $n_{p_i}=1\ \forall i$	n_{p_i} arbitrary
expected waiting time $E(V_{q_i})$ resp. $E(V_q)$	$\frac{\rho_i}{\mu_i(1-\rho_i)}$	$\frac{1}{\lambda_i^a+\lambda_i^s}\cdot\frac{\rho_i^{n_{p_i}+1}}{(n_{p_i}-1)!(n_{p_i}-\rho_i)^2}p_{o_i}$	$\frac{1}{\lambda_o^a+\lambda_o^s}\sum_{i=1}^{N}\frac{\rho_i^2}{1-\rho_i}$	$\frac{1}{\lambda_o^a+\lambda_o^s}\sum_{i=1}^{N}\frac{\rho_i^{n_{p_i}+1}}{(n_{p_i}-1)!(n_{p_i}-\rho_i)^2}p_{o_i}$
expected indwelling time $E(V_i)$ resp. $E(V)$	$\frac{1}{\mu_i(1-\rho_i)}$	$\frac{1}{\lambda_i^a+\lambda_i^s}\left(\rho_i+\frac{\rho_i^{n_{p_i}+1}}{(n_{p_i}-1)!(n_{p_i}-\rho_i)^2}\cdot p_{o_i}\right)$	$\frac{1}{\lambda_o^a+\lambda_o^s}\sum_{i=1}^{N}\frac{\rho_i}{1-\rho_i}$	$\frac{1}{\lambda_o^a+\lambda_o^s}\sum_{i=1}^{N}\left(\rho_i+\frac{\rho_i^{n_{p_i}+1}}{(n_{p_i}-1)!(n_{p_i}-\rho_i)^2}p_{o_i}\right)$
expected number of waiting patients $E(L_{q_i})$ resp. $E(L_q)$	$\frac{\rho_i^2}{1-\rho_i}$	$\frac{\rho_i^{n_{p_i}+1}}{(n_{p_i}-1)!(n_{p_i}-\rho_i)^2}p_{o_i}$	$\sum_{i=1}^{N}\frac{\rho_i^2}{1-\rho_i}$	$\sum_{i=1}^{N}\frac{\rho_i^{n_{p_i}+1}}{(n_{p_i}-1)!(n_{p_i}-\rho_i)^2}p_{o_i}$
expected number of patients in the system $E(L_i)$ resp. $E(L)$	$\frac{\rho_i}{1-\rho_i}$	$\rho_i+\frac{\rho_i^{n_{p_i}+1}}{(n_{p_i}-1)!(n_{p_i}-\rho_i)^2}p_{o_i}$	$\sum_{i=1}^{N}\frac{\rho_i}{1-\rho_i}$	$\sum_{i=1}^{N}\rho_i+\frac{\rho_i^{n_{p_i}+1}}{(n_{p_i}-1)!(n_{p_i}-\rho_i)^2}p_{o_i}$
mean load degree ξ_i resp. ξ	ρ_i	$\frac{\rho_i}{n_{p_i}}$	$\frac{1}{N}\sum_{i=1}^{N}\rho_i$	$\frac{1}{N}\sum_{i=1}^{N}\frac{\rho_i}{n_{p_i}}$

$$\rho_i = \frac{\lambda_i^a+\lambda_i^s}{\mu_i} \qquad p_{o_i} = \frac{1}{\sum_{n_i=0}^{n_{p_i}-1}\frac{\rho_i^{n_i}}{n_i!}+\frac{\rho_i^{n_{p_i}}}{(n_{p_i}-1)!(n_{p_i}-1)}}$$

diagnosis rooms of every service unit is arbitrary (n_{p_i} arbitrary) or whether every service unit consists of only one diagnosis room ($n_{p_i}=1$ for all i=1, . . . , N). As for the derivation of the indices one may be directed to the references Little (1961), Ferschl (1964), Gross/Harris (1974), Lemoine (1977), and Hegemann (1985).

3. The diagnosis system of the General Hospital in Hagen City

3.1 Structure, operating parameters and indices

The diagnostic system of the General Hospital in Hagen City comprises five departments: function diagnostics, sonography, endoscopy, computer tomography, and X-ray diagnostics, which are divided into twelve service units with one diagnosis room each. Under queuing theoretical aspects it is an M/M/12/∞/FCFS/2-model whose structure referring to number and designation of the service units, their services, staff and apparatus is shown in Table 2. The relevant examination times in the service units are between 8:00 A.M. and 4:00 P.M. with the X-ray diagnostics being manned 24 hours a day by a medical-technical assistant at least, though.

The parameters of the arrival processes of the out-patients and in-patients at the twelve diagnostic service units - important for the process structure of the diagnosis system - as well as the distribution of examination times in the service units and the routing matrices of the transition probabilities have been investigated within a three-week-period on the basis of diagnosis processes of more than 1,700 out-patients and in-patients by means of a self-recording-method of the medical personnel working in the diagnostic system. Table 3 contains the intensities of the arrival flows and the examination rates. Hereby the outside of the diagnostic system is called 0. Then the routing matrices R^a and R^s follow. From these quantities one can calculate the indices of the single service units and the total diagnosis system according to the relations shown in Table 1; they are compiled in Table 4. Among these indices the expected waiting times of the patients $E(V_{q_i})$ and $E(V_q)$ as well as the load degrees ξ_i and ξ are of special interest for the further considerations.

3.2 Weak point analysis of the diagnosis system

Looking at the indices listed up in Table 4 the great differences in the load degree ξ_i of the single diagnostic rooms come up as the first weak point of the diagnosis system of the General Hospital in Hagen. Whereas the load degree of the computerized tomography reaches about 70% and is therefore rather high, the load degrees of the X-ray rooms 2, 5.2 and 6 reaching 13.7%, 25.5% and 6.2% as well as of the sonography reaching 13.6% must be called very low. Furthermore, the considerable differences between the X-ray rooms 1 and 2 equipped nearly identically are especially remarkable.

High divergencies can also be stated for the patients' waiting times $E(V_{q_i})$. A patient in front of the computerized tomography and X-ray room 4 must wait for his examination up to one hour on average, whereas the examination starts mostly immediately in the

Table 2: Structure of the diagnosis system

area	room	services	executive personnel	important equipment
function diagnostics	function diagnostics 2 (FCT 2)	long-term-ECG resting-ECG rhythm-ECG pacemaker check phonocardiogram apexcardiogram carotid pulse curve orthostasis-ECG oscillogram	1 or 2 medical-technical assistants (and 1 doctor)	1 long-term ECG computer 1 ECG apparatus 1 ultrasonic doppler apparatus
	function diagnostics 3 (FCT 3)	exercise-ECG echocardiogram pulmonary function test dye-dilution	1 medical-technical assistant and 1 doctor	1 ECG apparatus 1 echocardiograph 1 spirometre 1 dye-dilution apparatus
sonography	sonography (SONO)	sonographic examination of the abdominal organs	1 doctor	1 sonographic equipment
endoscopy	endoscopy (ENDO)	gastroscopy colonoscopy ERCP esophageal dilation organ biopsy laparoscopy endoscopic sclerosizing of varices rectoscopy proctoscopy	1 or 2 medical-technical assistants and 1 or 2 doctors	1 operating table and different endoscopes
computerized tomography	computerized tomography (COMP)	abdominal CT spinal CT cerebral CT pancreatic CT thoratic CT kidney CT liver CT pelvic CT	2 medical-technical assistants (and 1 doctor)	1 computer tomograph
X-ray diagnostics	X-ray diagnostics 1 (X 1)	fluoroscopy and roentgenogram of all parts of the body upper gastro-intestinal series barium enema phlebography esophagus ERCP fistulography	1 medical-technical assistant and 1 or 2 doctors	1 Bucky table

Table 2 (continued)

X-ray diagnostics	X-ray diagnostics 2 (X 2)	arthrography myelography (otherwise as in X 1 but without phlebography)	1 medical-technical assistant and 1 doctor	1 Bucky table
	X-ray diagnostics 3 (X 3)	chest X-ray (2 levels) trachea cervical spine paranasal sinuses	2 medical-technical assistants	1 Bucky wall tripod
		tomography of all regions of the body drip infusion cholecystography drip infusion pyelography		1 Bucky table
		panoramic picture of the upper and lower jaw bones		1 status-X-apparatus
	X-ray diagnostics 4 (X 4)	renovasography arteriography of of arms and legs lymphangiography aortography	1 medical-technical assistant and 2 doctors	1 AOT changer
	X-ray diagnostics 5.1 (X 5.1)	bones cholecystography after intravenous injection thorax (inclined)	1 medical-technical assistant	1 Bucky table
		chest X-ray (2 levels) paranasal sinuses abdomen trachea spine (up-right-position)		1 Bucky wall tripod
	X-ray diagnostics 5.2 (X 5.2)	bones pyelography by intravenous injection function of articulations (knee or ankle) *Scheuba*	1 medical-technical assistant	1 Bucky table
	X-ray diagnostics 6 (X 6)	mammography pneumocystography galactography roentgenogram of the soft tissues of the hands	1 medical-technical assistant	1 mammomat

Table 3: Arrival and examination rates of the diagnostics

room	i	λ_i^a	λ_i^s	λ_i	μ_i
Outside of diagnostics	0	0,089444	0,153750	0,243194	-
FCT2	1	0,017083	0,065694	0,082777	0,193742
FCT3	2	0,007639	0,008889	0,016528	0,058173
SONO	3	0,001806	0,013611	0,015417	0,113483
ENDO	4	0,005357	0,010863	0,016220	0,042030
COMP	5	0,015667	0,010111	0,025778	0,037263
X 1	6	0,003889	0,013889	0,017778	0,037890
X 2	7	0,001667	0,001111	0,002778	0,020243
X 3	8	0,019722	0,050417	0,070139	0,207512
X 4	9	0,002222	0,000278	0,002500	0,008520
X 5.1	10	0,025417	0,032917	0,058334	0,112965
X 5.2	11	0,004861	0,005556	0,010417	0,040794
X 6	12	0,003472	0,000694	0,004166	0,066869

$R^a = (r^a_{ij}) =$

0	0,185	0,005	0,014	0,060	0,172	0,033	0,019	0,160	0,025	0,254	0,054	0,019
0,472	0	0,415	0	0	0	0	0	0,106	0	0,008	0	0
0,764	0	0	0	0	0	0	0	0,200	0	0,036	0	0
1,000	0	0	0	0	0	0	0	0	0	0	0	0
0,949	0	0	0,051	0	0	0	0	0	0	0	0	0
0,947	0,009	0	0	0	0	0	0	0,044	0	0	0	0
1,000	0	0	0	0	0	0	0	0	0	0	0	0
0,917	0	0	0	0	0	0	0	0,083	0	0	0	0
0,775	0,021	0,007	0,007	0	0,014	0,021	0	0	0	0,063	0	0,092
1,000	0	0	0	0	0	0	0	0	0	0	0	0
0,951	0	0	0,005	0	0	0,022	0	0,022	0	0	0	0
0,914	0	0	0	0	0	0	0	0,086	0	0	0	0
0,640	0	0	0	0	0	0	0	0,080	0	0,280	0	0

$$R^S=(r^S_{ij})=\begin{bmatrix}
0 & 0,232 & 0,033 & 0,065 & 0,063 & 0,050 & 0,060 & 0,005 & 0,261 & 0,001 & 0,196 & 0,032 & 0,002 \\
0,841 & 0 & 0,049 & 0,011 & 0,002 & 0,013 & 0,002 & 0,002 & 0,061 & 0 & 0,015 & 0,004 & 0 \\
0,953 & 0 & 0 & 0 & 0 & 0 & 0,016 & 0 & 0,031 & 0 & 0 & 0 & 0 \\
0,735 & 0,133 & 0 & 0 & 0,031 & 0,010 & 0,031 & 0 & 0,031 & 0 & 0,020 & 0,010 & 0 \\
0,858 & 0,077 & 0,013 & 0 & 0 & 0,013 & 0,013 & 0 & 0,013 & 0 & 0,013 & 0 & 0 \\
0,904 & 0,055 & 0,027 & 0 & 0 & 0 & 0 & 0 & 0 & 0 & 0,014 & 0 & 0 \\
0,810 & 0,070 & 0 & 0,010 & 0 & 0 & 0 & 0,010 & 0,090 & 0 & 0,010 & 0 & 0 \\
1,000 & 0 & 0 & 0 & 0 & 0 & 0 & 0 & 0 & 0 & 0 & 0 & 0 \\
0,452 & 0,419 & 0 & 0,052 & 0,008 & 0,011 & 0,022 & 0 & 0 & 0,003 & 0,019 & 0,006 & 0,008 \\
1,000 & 0 & 0 & 0 & 0 & 0 & 0 & 0 & 0 & 0 & 0 & 0 & 0 \\
0,654 & 0,131 & 0,004 & 0,004 & 0 & 0,021 & 0,076 & 0,004 & 0,105 & 0 & 0 & 0 & 0 \\
0,750 & 0,050 & 0 & 0 & 0,025 & 0 & 0,025 & 0 & 0,125 & 0 & 0,025 & 0 & 0 \\
0,600 & 0,200 & 0 & 0 & 0 & 0,200 & 0 & 0 & 0 & 0 & 0 & 0 & 0
\end{bmatrix}$$

Table 4: Indices of the single diagnostic service units and the total system

room	i	$E(V_{q_i})$ [min.]	$E(V_i)$ [min.]	$E(L_{q_i})$ [pat./min.]	$E(L_i)$ [pat./min.]	ξ_i [%]
FCT2	1	3,85	9,01	0,32	0,75	42,7
FCT3	2	6,82	24,01	0,11	0,40	28,4
SONO	3	1,38	10,19	0,02	0,16	13,6
ENDO	4	14,95	38,74	0,24	0,63	38,6
COMP	5	60,23	87,06	1,55	2,24	69,2
X 1	6	23,32	49,72	0,41	0,88	46,9
X 2	7	7,85	57,25	0,02	0,16	13,7
X 3	8	2,46	7,27	0,17	0,51	33,8
X 4	9	48,73	166,11	0,12	0,42	29,3
X 5.1	10	9,45	18,30	0,55	1,07	51,6
X 5.2	11	8,40	32,91	0.09	0,34	25,5
X 6	12	0,99	15,94	0,01	0,07	6,2
total system		$E(V_q)$ 14,9	$E(V)$ 31,3	$E(L_q)$ 3,6	$E(L)$ 7,6	ξ 33,3

sonography, in the X-ray rooms 3 and 6 and in the function diagnosis room 2.

The analysis of the indwelling times $E(V_i)$ shows extremely high values for the computerized tomography and the X-ray rooms 1, 2 and 4. In the X-ray rooms the long indwelling times must be attributed to the relatively long examination times, whereas the long indwelling time of nearly 1 1/2 hours in the computerized tomography can be explained by the large part of waiting times.

Comparing the waiting times with load degrees one meets the well-known dilemma of scheduling in the computerized tomography: the relatively high waiting times definitely result from the rather high load degree of nearly 70%. In the X-ray rooms 5.1 and 3 with comparatively higher load degrees only short waiting times appear because here examinations rarely last over a longer period. In the opposite sense, the dilemma of scheduling comes up in the sonography and in the X-ray room 6, too: the negligible short waiting times here solely result from the low load degrees and thus high dead times.

These weak points to be found mainly in the X-ray department can be eliminated partly by restructuring measures concerning the pooling of rooms and the shifting of staff as will be shown by the discussion of the rationalizing potential of the diagnostic system in the following chapter. The aim will be to reduce the great differences in the load degrees and to shorten the long waiting times by changing the capacity structure.

Concerning the diagnostic system as a whole the average waiting and indwelling times of the patients $E(V_q)$ and $E(V)$ can be considered as reasonable. The expected load degree ξ and thus the efficiency of the existing diagnosis system reaching 33.3% must be regarded as very low. Frequently, it is the aim of restructuring and rationalizing measures to strive for an average load degree of about 50%.

4. Possibilities of rationalization in the diagnosis system of the General Hospital in Hagen

4.1 Objectives

Generally, the object of the task to rationalize is the question: which modifying measures shall be carried out in which diagnosis rooms to which extent so that possibly high, steady load degrees and possibly low, reasonable waiting times for patients can be reached in the diagnosis system and in each diagnosis room? This limits the consideration to two system indices. There is a number of possibilities to operationalize this goal conflict of scheduling, which is a two-criteria decision problem, by formulating objective functions and constraints. In this paper the procedure to formulate the problem is as follows.

M may denote the set of all combinations m of improving measures concerning the single diagnostic service units, i.e.

$$M = \{m \mid m = (m_1, \ . \ . \ . \ , m_{12})\}.$$

From the point of view of hospital doctors and patients that alternative $m' M$ would be favored, for which the expected waiting time of patients $E(V_q)$ in the total diagnosis system becomes a minimum, i.e.

$$E(V_q(m^*)) = \min_{m \in M} \left\{ E(V_q(m)) = \frac{1}{\lambda_0^a + \lambda_0^s} \sum_{i=1}^{N(m)} \frac{\rho_i^2\ (m)}{1-\rho_i\ (m)} \right\}$$

must be fulfilled with $N(m) \leq 12$ denoting the number of diagnosis rooms possibly changed due to the pooling of rooms. This criterion tallies with the minimization of the average total time and the minimization of the expected number of patients in the diagnosis system.

On the contrary the alternative $\tilde{m}$ takes into account the objective of a possibly high load degree of the total diagnosis system usually important for the hospital management and the state. Its determination leads to the goal formulation

$$\xi(\tilde{m}) = \max_{m \in M} \left\{ \xi(m) = \frac{1}{N(m)} \sum_{i=1}^{N(m)} \rho_i(m) \right\}.$$

This is identical with the minimization of dead times and analogous to the minimization of costs, when the service spectrum is given.

Considering these criteria for the total system it must be guaranteed, however, that the patients' waiting times in front of every diagnosis room do not exceed a reasonable limit $\bar{W}$ which is ordinarily rated at $\bar{W} = 30$ minutes by the public health policy. Therefore the feasible alternatives $m \in M$ are subject to the restrictions

$$E(V_{q_i}(m)) = \frac{\rho_i(m)}{\mu_i(m)\,(1-\rho_i(m))} \leq \bar{W},\ i=1, \ldots, N(m),\ m \in M.$$

Further constraints resulting from medical needs, room situations and technical border conditions may be added. They shall not be discussed here any further, but will be implied by the definition of the measures in the following chapter for reasons of simplicity. So the problem of rationalization may be described by the decision problem or vector minimization problem, respectively with

$$\text{"}\min_{m\in\hat{M}}\text{"} \quad \begin{pmatrix} E(V_q(m)) \\ -\xi(m) \end{pmatrix}$$

with

$$\hat{M} = \{m\in M \mid E(V_{q_i}(m)) \leq 30 \text{ for all } i=1, . . ., N(m)\}.$$

Efficient compromise measures have to be taken into account as solutions not dominated by any other rationalizing measure referring to the waiting times of patients and the load degree of the total diagnosis system.

4.2 Rationalizing measures and their results

Possible rationalizing measures are the removal of examination services into other diagnostic service units, the reinstallation of equipment and the shifting of staff as well as the fusion of examination rooms, the new purchase of more powerful examination apparatus and the sacking or employment of staff. Such measures effect the number of diagnosis rooms, the arrival and examination rates of the service units, the structure of the routing matrices and, furthermore, the average waiting times and load degrees of the diagnosis rooms and the total diagnosis system. Correspondingly, it will be important to point out alternatives of the capacity structure of the diagnosis system of the General Hospital in Hagen possibly leading to more favorable constellations of the queuing theoretic indices than at present.

Concerning their temporal possibilities of realization the measures of improvement to be considered must be differentiated according to their appointed times. Whereas the removal of examination services, the reinstallation of equipment, the shifting and new employment of staff can be carried out within a medium-term period, the fusion of rooms, the purchase of equipment, and the sacking of staff can only be done within a long-term period. As for the rest, attention was directed to the definition of such rationalizing measures to be mastered mainly with the present equipment and personnel and not leading to the sacking of staff within a medium-term period.

The possible measures of an organizational restructuring first concerning single diagnostic rooms, only, shall formally be described by components. A total measure m then results from the componentwise combination of single measure which can be planned for the different diagnostic service units. As the number of restructuring measures for the diagnosis system discussed is finite, the vector minimization problem, which was formulated in the last chapter, is a discrete optimizing problem. In order to make this more concrete

$$m_{ij}^{\nu} = (m_i^{\nu}, m_j^{\nu}),\ \nu \in \mathbb{N},\ i,\ j \in \{1, . . . ,12\},\ i \neq j,$$

may denote a component-wise single measure ν referring simultaneously to the diagnosis rooms i and j. In case of such a single measure referring to one diagnostic service unit only the notation with vector components

$$m_i^{\nu},\ \nu \in \mathbb{N},\ i \in \{1, . . . , 12\}$$

is sufficient. It is clear that changing nothing is a single measure as well. This fact may be described by

$$m_{ij}^0 = (m_i^0, m_j^0) \text{ or } m_i^0, \text{ respectively.}$$

The actual state of the diagnosis system corresponds to the total measure

$$m^0 = (m_1^0, \ldots, m_{12}^0).$$

The single measures of rationalization listed up in the following are - due to the preceding weak point analysis - directed to reduce the waiting times of patients at the single diagnostic service units below the reasonable level and, at the same time, not losing sight of a more equal and better load degree of the diagnosis rooms. For medical-technical reasons there are no possibilities to change anything in the function diagnostics, sonography, and endoscopy so that for these rooms every total measure is characterized by

$$m_i = m_i^0 \text{ for } i=1, \ldots, 4.$$

Consequently, objects for the rationalizing measures are the computerized tomography and the service units of the X-ray department. The following measures may be considered for them:

m_{67}^0: The situation in the X-ray rooms 1 and 2 (service units 6 and 7) remains unchanged.

m_{67}^1: Apart from the phlebographies and fistulographies all fluoroscopies and roentgenograms of all parts of the body as well as the examinations upper gastro-intestinal series and barium enema are carried out in the X-ray room 1. In the X-ray room 2 the diagnostic measures esophagus and ERCP are carried out as well as arthrographies and myelographies.

m_{67}^{2}: With the exception of fluoroscopies and roentgenographies, phlebographies and barium enema taking place in the X-ray room 1 all other diagnostics examinations are solely carried out in the X-ray room 2.

m_{67}^{3}: Unlike m_{67}^{2} the ERCP examinations are to be offered not in the X-ray room 2, but in the X-ray room 1.

m_{67}^{4}: Pooling of X-ray rooms 1 and 2 with the intention to purchase a more powerful fluoroscopy apparatus which will help to offer the full spectrum of the services of the both existing rooms and will be operated by one medical-technical assistant only.

m_{67}^{5}: Unlike m_{67}^{4} two medical-technical assistants will be charged with the carrying out of the diagnostic services in the new X-ray room 1/2 ([6+7]).

$m_{8,12}^{0}$: The situation in the X-ray rooms 3 and 6 (service units 8 and 12) remains unchanged.

$m_{8,12}^{6}$: The X-ray rooms 3 and 6 are merged ([8+12]) with the intention to install the mammomat of X-ray room 6 in the X-ray room 3. The number of staff in the X-ray room 3 remains unchanged, though.

m_{9}^{0}: The situation in the X-ray room 4 (service unit 9) remains unchanged.

m_{9}^{7} Direct preparation times needed for the examinations in the X-ray room 4 are to be transferred partially to the beginning of the examination day.

m_{9}^{8}: In addition to m_{9}^{7} replacement of the old surgical equipment by a powerful new apparatus.

$m^0_{10,11}$: The situation in the X-ray rooms 5.1 and 5.2 (service units 10 and 11) remains unchanged.

$m^9_{10,11}$: Fusion of the X-ray rooms 5.1 and 5.2 ([10+11]) by dis-investment of the Bucky table in room 5.2 and removing the lead wall. As formerly the staff will consist of two medical-technical assistants.

$m^{10}_{10,11}$: Similar to $m^9_{10,11}$; but only one medical-technical assistant will be assigned to the combined diagnosis room [10+11].

m^0_5: The situation in the computerized tomography (service unit 5) remains unchanged.

m^{11}_5: The staff in the computerized tomography will be enlarged to three jobs for medical-technical assistants.

The single measures mentioned are compiled in Table 5 as to their effects of the service units the patients' waiting times and the load degrees. The letters mt and lt in the second column of Table 5 indicate the medium-term or long-term character of the measures whereby every medium-term measure can also be realized in the long-term. It is conspicuous that the measures m^8_9 and m^{11}_5 could cause the urgently desired drastic reductions in the waiting times of patients especially in the X-ray room 4 (service unit 9) and in the computerized tomography (service unit 5). The measure $m^6_{8,12}$ causes a 100%-increase in the load degree of the combined service unit [8+12] with the patients' waiting times being negligibly longer.

Because of the independence premises (P 4) and (P 10) a total rationalizing measure in the diagnostic system consists of a combination of single measures for the different diagnostic rooms. Thereby it is reasonable that only the combination of single measures being themselves feasible and efficient can lead to a feasible and efficient total measure. The reverse is not valid, at least not as far as efficiency is concerned! In the first column of Table 5 the feasible and efficient single measures are marked by little rectangles. The distinct dominance of the medium-term measure m_{67}^{2} over the measures m_{67}^{0}, m_{67}^{1}, m_{67}^{3} is dependent on the fact that in case of the examination rooms 6 and 7 being regarded as one service unit [6+7] the total waiting time turns out the shortest while the load degree remains unchanged. At the same time this is the reason for the dominance with respect to its contribution to the total waiting time and the total load degree of the diagnostic system (cf. formulas of the indices in Table 1), which are the criteria of the vector minimum problem.

The values in Table 4 and 5 elucidate at once that the actual state of the diagnosis system, i.e. the total measure m^{0} is not feasible and, additionally, there exists no feasible medium-term total measure m of improvement, because in both cases the waiting times in the service unit 9 (X-ray room 4) are always above 30 minutes. Remembering that $m_{i} = m_{i}^{0}$ must hold for the first four service units - these single measures are all feasible - and taking into account that all single measures feasible for the other service units are listed up in table 5 one receives by the combination of the feasible and efficient single measures 3x2x1x2x1=12 feasible, alternative, long-term total measures, which have to be checked according to their efficiency and

Table 5: Single measures and their effects

measure	medium-term (mt) or long-term (lt)	$E(V_{q_i})$ [min.]			ξ_i [%]		
		i=6	i=7	[6+7]	i=6	i=7	[6+7]
m^o_{67}	mt/lt	23,32	7,85	21,37	46,9	13,7	30,3
m^1_{67}	mt/lt	17,86	9,72	15,93	40,5	20,0	30,3
m^2_{67}	mt/lt	12,95	12,68	12,92	31,9	28,6	30,3
m^3_{67}	mt/lt	16,18	9,99	14,02	36,6	23,9	30,3
m^4_{67}	lt	-	-	25,13	-	-	50,4
m^5_{67}	lt	-	-	17,26	-	-	44,3
		i=8	i=12	[8+12]	i=8	i=12	[8+12]
$m^o_{8,12}$	mt/lt	2,46	0,99	2,41	33,8	6,2	20,0
$m^6_{8,12}$	mt/lt	-	-	3,72	-	-	40,0
				i=9			i=9
m^o_9	mt/lt			48,73			29,3
m^7_9	mt/lt			39,39			26,8
m^8_9	lt			27,74			23,1
		i=10	i=11	[10+11]	i=10	i=11	[10+11]
$m^o_{10,11}$	mt/lt	9,45	8,40	9,32	51,6	25,5	38,6
$m^9_{10,11}$	lt	-	-	10,74	-	-	56,6
$m^{10}_{10,11}$	lt	-	-	37,98	-	-	77,2
				i=5			i=5
m^o_5	mt/lt			60,23			69,2
m^{11}_5	mt/lt			28,52			56,5

optimality. They are compiled in Table 6 as the measures m^{101} through m^{112}. To make comparisons possible the actual state m^0 and a medium-term total measure $\hat{m}$ (both unfeasible) are listed up as well. One sees, that not all but five long-term measures are efficient with regard to the total waiting time of patients $E(V_q)$ and the total load degree ξ of the diagnosis system. On the other hand, there exist nine feasible long-term total measures dominating the actual state of the diagnosis system at the General Hospital in Hagen. The unfeasible medium-term total measure $\hat{m}$ also dominates the actual state and is efficient even in comparison with the long-term measures. A number of measures leads to the reduction in the number N(m) of examination rooms with the spectrum of services unchanged and - at the same time - to an improved service quality and load degree of the total diagnosis system.

Measures with the component m_{67}^2 clearly show a lower load degree of the service units 6 and 7 and and a lower total load. Those with the component m_{67}^4 unsatisfactorily increase the waiting times of patients in front of the X-ray rooms 1 and 2, if they were considered as a unit, in comparison with the actual state. Therefore, the measure m^{112} seems to be the optimal compromise within the long-term period for the doctors' working in the diagnosis system. Compared with the actual state this measure increases the total load of the diagnosis system by 18% and decreases the average patients' waiting times by 20%. It was impossible to reach a total load degree of 50% by the measures discussed.

Table 6: Rationalizing measures, their characteristics and results

	actual state	medium-term	long-term											
	m^{o}	$\hat{m}$	m^{1o1}	m^{1o2}	m^{1o3}	m^{1o4}	m^{1o5}	m^{1o6}	m^{1o7}	m^{1o8}	m^{1o9}	m^{11o}	m^{111}	m^{112}
MEASURES	m_1^{o}	m_1^{o}	m_1^{o}	m_1^{o}	m_1^{o}	m_1^{o}	m_1^{o}	m_1^{o}	m_1^{o}	m_1^{o}	m_1^{o}	m_1^{o}	m_1^{o}	m_1^{o}
	m_2^{o}	m_2^{o}	m_2^{o}	m_2^{o}	m_2^{o}	m_2^{o}	m_2^{o}	m_2^{o}	m_2^{o}	m_2^{o}	m_2^{o}	m_2^{o}	m_2^{o}	m_2^{o}
	m_3^{o}	m_3^{o}	m_3^{o}	m_3^{o}	m_3^{o}	m_3^{o}	m_3^{o}	m_3^{o}	m_3^{o}	m_3^{o}	m_3^{o}	m_3^{o}	m_3^{o}	m_3^{o}
	m_4^{o}	m_4^{o}	m_4^{o}	m_4^{o}	m_4^{o}	m_4^{o}	m_4^{o}	m_4^{o}	m_4^{o}	m_4^{o}	m_4^{o}	m_4^{o}	m_4^{o}	m_4^{o}
	m_5^{o}	m_5^{11}	m_5^{11}	m_5^{11}	m_5^{11}	m_5^{11}	m_5^{11}	m_5^{11}	m_5^{11}	m_5^{11}	m_5^{11}	m_5^{11}	m_5^{11}	m_5^{11}
	m_6^{o}	m_6^{2}	m_6^{2}	m_6^{2}	m_6^{2}	m_6^{2}	m_6^{4}	m_6^{4}	m_6^{4}	m_6^{4}	m_6^{5}	m_6^{5}	m_6^{5}	m_6^{5}
	m_7^{o}	m_7^{2}	m_7^{2}	m_7^{2}	m_7^{2}	m_7^{2}	m_7^{4}	m_7^{4}	m_7^{4}	m_7^{4}	m_7^{5}	m_7^{5}	m_7^{5}	m_7^{5}
	m_8^{o}	m_8^{6}	m_8^{o}	m_8^{o}	m_8^{6}	m_8^{6}	m_8^{o}	m_8^{o}	m_8^{6}	m_8^{6}	m_8^{o}	m_8^{o}	m_8^{6}	m_8^{6}
	m_9^{o}	m_9^{7}	m_9^{8}	m_9^{8}	m_9^{8}	m_9^{8}	m_9^{8}	m_9^{8}	m_9^{8}	m_9^{8}	m_9^{8}	m_9^{8}	m_9^{8}	m_9^{8}
	m_{1o}^{o}	m_{1o}^{o}	m_{1o}^{o}	m_{1o}^{9}	m_{1o}^{o}	m_{1o}^{9}	m_{1o}^{o}	m_{1o}^{9}	m_{1o}^{o}	m_{1o}^{9}	m_{1o}^{o}	m_{1o}^{9}	m_{1o}^{o}	m_{1o}^{9}
	m_{11}^{o}	m_{11}^{o}	m_{11}^{o}	m_{11}^{9}	m_{11}^{o}	m_{11}^{9}	m_{11}^{o}	m_{11}^{9}	m_{11}^{o}	m_{11}^{9}	m_{11}^{o}	m_{11}^{9}	m_{11}^{o}	m_{11}^{9}
	m_{12}^{o}	m_{12}^{6}	m_{12}^{o}	m_{12}^{o}	m_{12}^{6}	m_{12}^{6}	m_{12}^{o}	m_{12}^{o}	m_{12}^{6}	m_{12}^{6}	m_{12}^{o}	m_{12}^{o}	m_{12}^{6}	m_{12}^{6}
$E(V_q)$	14,9	11,1	1o,6	11,o	1o,9	11,4	11,6	12,1	12,o	12,4	1o,9	11,4	11,4	11,8
ξ	33,3	34,9	31,7	32,7	34,6	36,o	33,7	35,o	37,o	38,9	33,1	34,4	36,4	38,2
N(m)	12	11	12	11	11	1o	11	1o	1o	9	11	1o	1o	9
feasible	-	-	+	+	+	+	+	+	+	+	+	+	+	+
efficient	-	+	+	-	+	-	-	-	-	+	-		+	+

REFERENCES

1. Bailey, N.T.J. (1954). Queueing for Medical Care, in: Journal of the Royal Statistical Society 3,Series C, pp. 137-145.

2. Burke, P.J. (1956). The Output of a Queuing System,in:Operations Research 4, pp. 699-704.

3. Burke, P.J. (1972). Output Processes and Tandem Queues, in: Proceedings of the Symposium on Computer - Communications Networks and Teletraffic, New York, pp. 419-428.

4. Carruthers, M.E. (1970). Computer Analysis of Routine Pathology Work Schedules Using a Simulation Programme,in: Journal of Clinical Pathology 23, pp. 269-272.

5. Eichhorn, S. (1976). Krankenhausbetriebslehre, Vol. II, 3rd edition, Köln.

6. Ferschl, F. (1964). Zufallsabhängige Wirtschaftsprozesse, Wien/ Würzburg.

7. Fleck, H. (1977). Analyse der Ablauforganisation in einem medizinischen Untersuchungszentrum mit Hilfe der Netzwerktechnik,in: Methods of Information in Medicine, 16, pp. 81-88.

8. Gierl, L. (1976). Partialmodelle der Ablaufplanung im Krankenhaus auf empirischer Grundlage, Ph.D. thesis, Nürnberg.

9. Gross, D. and Harris, C.M. (1974). Fundamentals of Queuing Theory, New York/London/Sydney/Toronto.

10. Hegemann, H. (1985). Zur Kapazitäts- und Prozeßplanung in der klinischen Diagnostik, Berlin/Heidelberg/New York/Tokyo.

11. Jackson, J.R. (1963). Jobshop-like Queuing Systems, in: Management Science 10, pp. 131-142.

12. Kendall, D.G. (1953). Stochastic Processes Occuring in the Theory of Queues and their Analysis by the Method of the Imbedded Markov Chain,in:The Annals of Mathematical Statistics 24, pp. 338-354.

13. Klimov, G.P. (1979). Bedienungsprozesse, Basel/Stuttgart.

14. Lemoine, A.J. (1977). Networks of Queues - A survey of Equilibrium Analysis,in: Management Science 24, pp. 464-481.

15. Little, J.D.C. (1961). A Proof of the Queueing Formula $L = \lambda W$, in: Operations Research, 9, pp. 383-387.

16. Melamed, B. (1979a). On Poisson Traffic Processes in Discrete-State Markovian Systems with Applications to Queueing Theory, in: Advances in Applied Probability, 11, pp. 218-239.

17. Melamed, B. (1979b). Characterizations of Poisson Traffic Streams in Jackson Queuing Networks,in: Advances in Applied Probability 11, pp. 422-438.

18. Neuman, K. and Steinhardt, U. (1979). GERT Networks and the Time-Oriented Evaluation of Projects,in: Lecture Notes in Economics and Mathematical Systems, Vol. 172, Berlin/ Heidelberg/New York.

19. O'Kane, P.C. (1981). A Simulation Model of a Diagnostic Radiology Department,in: European Journal of Operational Research 6, pp. 38-45.

20. Pritsker, A.A.B. and Happ, W.W. (1966). GERT: Graphical Evaluation and Review Technique,in: Journal of Industrial Engineering 17, pp. 267-274 and pp. 293-301.

21. Revesz, G., Shea, F.J. and Ziskin, M.C. (1972). Patient Flow and Utilization of Resources in a Diagnostic Radiology Department, in:Radiology, 102, pp. 21-30.

22. Taylor, B.W. and Keown, A.J. (1980). A Network Analysis of an Inpatient-Outpatient Department,in: Journal of the Operational Research Society, 31, pp. 169-179.

23. Taylor, I.D.S. and Templeton, J.G.C. (1980). Waiting Time in a Multi-Server Cutoff-Priority Queue, and its Application to an Urban Ambulance Service,in: Operations Research 28, pp. 1168-1188.

24. Walrand, J. (1982). On the Equivalence of Flows in Networks of Queues,in: Journal of Applied Probability 19, pp. 195-203.

MULTIPLE CRITERIA DECISION SUPPORT SYSTEMS IN PRODUCTION PLANNING: A MICRO EXPERIENCE

Birsen Karpak
Department of Management
Youngstown State University
Youngstown, OH 44555

ABSTRACT

The implementation of the modified two person zero-sum game approach, contracting cone method and augmented Tchebycheff method on a production planning problem in a manufacturing organization is explained. Karpak, Kuruuzum (1987) compared the modified two person zero-sum game approach with the contracting cone method on a mainframe. In this study we describe our experience on a microcomputer.

1. INTRODUCTION

This study is an extension of author's earlier joint work (Karpak, Kuruuzum 1987). Applications of multiple criteria decision making methods are very few despite the abundance of algorithms developed. Since we believe that these techniques will be much more extensively used if they are available on microcomputers, we experimented on them using an IBM-PC XT.

Among the existing algorithms we have selected the augmented Tchebycheff method (Steuer, 1986 pp. 419 - 450) since it produces efficient extreme as well as nonextreme points and it can be solved together with any commercial linear programming Software.

NATO ASI Series, Vol. F56
Multiple Criteria Decision Making and Risk Analysis
Using Microcomputers. Edited by B. Karpak and S. Zionts

The other two methods we have used are the modified two person zero-sum method (Karpak, 1978) and the contracting cone method (Steuer 1977 and 1986). We had favorable experience with last two on a mainframe (Karpak, Kuruuzum 1987). However, with the contracting cone method, one can only produce extreme efficient points.

As software we used VINO (Cunningham, Schrage 1987). Data entry is quite facilitated in LOTUS based VINO. There are two versions: standard VINO requires 448 K memory whereas super VINO requires 640 K. We worked with the standard version. Maximum allowable numeric cells are 3000. Although it is theoretically possible to solve problems with 750 decision variables and 750 constraints, the size of the problem can be solved diminishes remarkably if the technological matrix is not sparse.

The paper is organized as follows: section 2 describes the production system. In section 3 a production planning with the contracting cone method is given. Section 4 illustrates the results obtained with the augmented Tchebycheff. Implementation of the modified two person zero-sum approach is given in section 5. Section 6 contains a summary and discussion.

2. PRODUCTION SYSTEM

The production system considered is quite similar to the one in our previous study. However the portfolio of resources has changed. As a result of our former study, the company has adjusted the capacity of assembly line II. This had an impact on the resources as well as on the objective functions. Another major change was in the quality control department. When we revisited the company,we saw that the bottleneck on quality control department could be released. We used the idea of Zeleny. Rather than optimizing the original system, we redesigned the system and then optimized it(Zeleny,1987). This redesign was a result of this study. In the new model of the system we have also treated the marketing constraints differently. Because of all these changes and for the sake of traceability we shall briefly redescribe the entire system. The interested reader is referred to the unabridged version of this paper which may be obtained from the author.

The company manufactures 96 different products. They may be grouped as D type fuses, HRC fuse cartridges, HRC load switches, cylindrical fuse cartridges, porcelain lampholders, fixation materials, ceramic insulators and English type fuses.

The new portfolio of resources, usage of the resources by each product and sales forecasts for each product are given in unabridged version of the paper (Karpak, 1988).

In our previous approach, sales forecasts had been taken as the maximum amount marketable for each product. In this new model, when we solved the problem with sales forecasts as upperbounds, we found a considerable excess of many resources. We were able to produce a maximum of almost all products. Our problem seemed to be a marketing problem rather than a production planning problem. When we presented the results, management wanted us to find the most profitable products. They were ready to increase marketing efforts for those products produced in excess of the sales forecasts.

The company is primarily interested in profit maximization. However, they found too much idle capacity in certain resources when maximizing profit. In order to be able to use these resources, the glazing and assembly line departments had to work overtime. Management wanted to minimize overtime. They were eager to keep manufacturing costs as low as possible. This time, the conflicting objectives of the system are:

1. Maximization of profit
2. Minimization of manufacturing cost
3. Minimization of overtime in the glazing department
4. Minimization of overtime in the assembly departments

3. PRODUCTION PLANNING WITH THE CONTRACTING CONE METHOD

The essence of the contracting cone method(CCM) is reducing the criterion cone at each iteration. Assuming we have k objective functions, initially the method generates 2k + 1 nondominated extreme point solutions corresponding the following weighting vectors:

$\lambda_1 = (1,0,...,0)$
$\lambda_2 = (0,1,...,0)$
.
.
$\lambda_k = (0,0,...,1)$
$\lambda_{k+1}=(1/k^2,r,r,...,r)$
$\lambda_{k+2}=(r,1/k^2,r,...,r)$
.
.
$\lambda_{k+k}=(r,r,...,r,1/k^2)$
$\lambda_{2k+1}=(1/k,1/k,...,1/k)$

where $r = (k+1)/k^2$

For each weighting vector λ, the $\lambda' Cx$ linear programming problem is maximized and the decision maker is presented with

2k+1 solutions. In the next iteration the volume of the cone is reduced by 1/k around the solution selected by the decision maker. If the decision maker selects the solution corresponding to one of the first k λ, for example λ_i, where $i \le k$, new λ will be,

$$\begin{array}{l} \quad\quad\quad\quad i^{th} \\ \quad\quad\quad\quad \downarrow \\ (0,0,....,1,...,0) \\ (v,0,....,p,...,0) \\ (0,v,....,p,...,0) \\ \quad\quad \cdot \\ \quad\quad \cdot \\ (0,0,....,p,...,v) \end{array}$$

where $v = k^{-1/(k-1)}$ and $p = 1-v$

If the decision maker selects the solution corresponding to λ_{k+i} where $i \le k$, new λ will be

$$\begin{array}{l} \quad\quad\quad\quad i^{th} \\ \quad\quad\quad\quad \downarrow \\ (t,t,...,v,t,....,t) \\ (t,t,...,0,t,..,t,u) \\ (t,t,...,0,t,..,u,t) \\ \quad\quad \cdot \\ \quad\quad \cdot \\ (u,t,...,0,t,....,t) \end{array}$$

where $t = p/(k-1)$ and $u = 1-(k-2)t$

If the decision maker selects the solution corresponding to λ_{2k+1} then new λ values will be

$$\begin{array}{l} (p/k,p/k,...,p/k,\ q\ ) \\ (p/k,...,p/k,\ q,\ p/k) \\ \quad\quad \cdot \\ \quad\quad \cdot \\ (\ q,\ p/k,p/k,...,p/k) \end{array}$$

where $q = 1-(k-1)p/k$

Interaction with the decision maker and contraction of the criterion cone continues until the decision maker is satisfied with a proposed solution.

First we solved the problem with the original 2k+1 λ's. Instead of 0, we have used $\epsilon > 0$ in order to eliminate dominated solutions. ϵ is taken as 0.0001. In our mainframe experiments we had to take $\epsilon = 0.01$. ADBASE (Steuer,1983) treated any $\epsilon \leq 0.001$ as $\epsilon = 0$. In other words we were not able to eliminate any dominated solutions. Table I gives the results obtained with this 2k+1 initial λ values.

TABLE I

Results Obtained with Initial 2k+1 λ values

	Profit (Turkish Lira) (millions)	Manufacturing Cost (Turkish Lira) (millions)	Overtime in Glazing (hours)	Overtime in Assembly Lines (hours)
λ_1	6,116	1,857	1,248	78,624
λ_2	2,209	696	0	0
λ_3	5,519	1,411	0	78,624
λ_4	3,894	1,115	552	0
λ_5	3,050	730	62	0
λ_6	6,086	1,878	552	78,624
λ_7	5,980	1,525	1,248	78,624
λ_8	5,934	1,521	552	78,624
λ_9	5,934	1,521	552	78,624

where

$\lambda_1 = (.9997, .0001, .0001, .0001)$
$\lambda_2 = (.0001, .9997, .0001, .0001)$
$\lambda_3 = (.0001, .0001, .9997, .0001)$
$\lambda_4 = (.0001, .0001, .0001, .9997)$
$\lambda_5 = (.0625, .3125, .3125, .3125)$
$\lambda_6 = (.3125, .0625, .3125, .3125)$
$\lambda_7 = (.3125, .3125, .0625, .3125)$
$\lambda_8 = (.3125, .3125, .3125, .0625)$
$\lambda_9 = (.25, .25, .25, .25)$

Of the solutions the decision maker selected the one corresponding to λ_4. Therefore we contracted the criterion cone around this point and solved the problem with the following first four λ values:

λ_1 = (.0001, .0001, .0001, .9997)
λ_2 = (.0001, .0001, .6298, .37)
λ_3 = (.0001, .6298, .0001, .37)
λ_4 = (.6298, .0001, .0001, .37)

Objective function values obtained with this contracted cone are given in table II

TABLE II

Objective Function Values With Contracted Gradient Cone

PROFIT (Turkish Lira) (millions)	Manufacturing Cost (Turkish Lira) (millions)	Overtime in Glazing (hours)	Overtime in Assembly Lines (hours)
3,894	1,115	552	0
3,688	1,186	0	0
2,209	696	0	0
6,116	1,857	1,248	78,624
4,752	1,111	1,173	30,552
6,111	1,783	672	70,523
5,789	1,843	1,248	1,250
5,481	1,750	1,248	78,624
4,980	1 854	552	0

Of the solutions our decision maker selected the solution corresponding to λ_5; i.e.the solution yielding a profit of 4.7 billion Turkish Lira, 1,173 hours overtime in glazing, and 30,552 hours overtime in assembly departments. The decision maker was satisfied with the solution. He did not want any further solutions.

4. RESULTS OBTAINED WITH INTERACTIVE TCHEBYCHEFF

We used the same λ values for interactive Tchebycheff method(ITM) as we used for CCM though Steuer recommends randomly generating certain number of λ and then selecting the most dispersed ones using the FILTER algorithm.(Steuer, 1986 p.448)

The ITM could briefly be explained as follows: Let J^{max} denote the indices of the maximization objectives and J^{min} the indices of the minimization objectives. Then the Tchebycheff first finds z* ideal criterion vector where

$$z_i^* = \begin{cases} \max\{c^i x + \epsilon_i \mid x \epsilon S\} & i \epsilon J^{max} \\ \min\{c^i x - \epsilon_i \mid x \epsilon S\} & i \epsilon J^{min} \end{cases}$$

The ϵ_i are numerically significant small positive numbers and S is the feasible region. Then we form the weighted Tchebycheff program(WTP)

$$\min \{ \alpha + \rho \sum_{i=1}^{k} w_i \}$$

$$\begin{array}{lll} \text{s.t.} & \alpha \geq \lambda_i w_i & 1 \leq i \leq k \\ & w_i = z_i^* - z_i & i \epsilon J^{max} \\ & w_i = z_i - z_i^* & i \epsilon J^{min} \\ & z_i = c^i x & 1 \leq i \leq k \\ & x \epsilon S & \end{array}$$

in which ρ is sufficiently small positive number. (See Silverman, Steuer, Whisman, 1987 p.120 and Steuer & Choo 1983). We generate initial 2k+1 nondominated solutions using the same λ's as we used for CCM method and then reducing the Λ space around the solution which our decision maker selected. We continue generating nondominated solutions until our decision maker is satisfied with one of them.

We have generated the initial 2k+1 solutions using the same initial 2k+1 λ values as we used in the CCM. Objective function values with these initial 2k+1 λ values are given in Table III.

TABLE III

Objective Function Values With Initial 2k+1 λ, Using Interactive Tchebycheff

PROFIT (Turkish Lira) (millions)	Manufacturing Cost (Turkish Lira) (millions)	Overtime in Glazing (hours)	Overtime in Assembly Lines (hours)
6,116	1,854	1,248	78,624
2,287	697	142	0
5,381	1,431	0	73,489
3,967	1,185	821	0
4,448	1,030	1,248	33,354
5,571	1,707	1,171	54,425
5,477	1,334	1,248	63,820
5,613	1,199	1,248	78,624
5,477	1,334	1,248	63,820

Of the solutions, the decision maker selected the one corresponding to λ_5. Therefore we reduced the weight space around this solution and found the following λ values:

$$\lambda_1 = (.62\ \ ,.1234,.1234,.1234)$$
$$\lambda_2 = (.0001,.1234,.1234,.7533)$$
$$\lambda_3 = (.0001,.1234,.7533,.1234)$$
$$\lambda_4 = (.0001,.7533,.1234,.1234)$$

Results corresponding these λ are given in table IV.

TABLE IV

Objective Function Values in Reduced Weight Space
Using Interactive Tchebycheff

PROFIT (Turkish Lira) (millions)	Manufacturing Cost (Turkish Lira) (millions)	Overtime in Glazing (hours)	Overtime in Assembly Lines (hours)*
5,958	1,508	1,248	78,624
2,485	699	29	-5,865
2,471	699	4	-6,296
2,280	697	31	-1,703
4,115	939	1,248	24,305
5,444	1,103	1,248	78,624
5,202	1,249	1,248	54,843
5,352	1,606	1,248	45,876
5,150	1,229	1,248	53,316

Of the solutions the decision maker selected the one corresponding to λ_5.; i.e. the solution yielding a profit of 4.1 billion Turkish Lira, 1,248 hours overtime in glazing, and 24,305 hours overtime in assembly. If one compares this decision with the one found by the CCM, both solutions could be considered close enough from decision maker's point of view.

5. RESULTS WITH MODIFIED TWO PERSON ZERO-SUM METHOD

In modified two person zero-sum, initially we optimize each objective function successively and find the values of others at this point after eliminating alternative solutions in case they exist (Karpak,Kuruuzum 1987). Entries of the pay-off matrix are normalized and an optimal weighting vector λ' is obtained by

*(-) sign in assembly overtime signifies slack capacity.

applying linear programming to this normalized pay-off matrix (Belenson,Kapur 1973). $\lambda^T Cx$ is maximized in feasible space to obtain nondominated solutions, where $n_i = \lambda'/M_i$, M_i is the maximum value of objective i and $\lambda_i = n_i/(\sum_{i=1}^{k} n_i)$.

The solution vector is x^*. If Cx^* is the optimum from decision maker's point of view the solution has been obtained; if not, unsatisfied objectives may be increased if the decision maker accepts some decrease in other objectives. In this decision the analyst may help the decision maker by giving the explicit trade-offs between objectives with the help of sensitivity analysis.

We optimized the objective functions successively and obtained the the pay-off table V after eliminating dominated solutions.

TABLE V

Solutions Obtained by Optimizing Each Objective Function

PROFIT (Turkish Lira) (millions)	Manufacturing Cost (Turkish Lira) (millions)	Overtime in Glazing (hours)	Overtime in Assembly Lines (hours)
6,116	1,854	1,248	78,624
2,287	696	142	0
5,381	1,431	0	73,489
3,967	1,185	821	0

We have normalized the coefficients of each objective function and then applied the modified two person zero-sum method to obtain the following λ values:

$\lambda_1^* = .04$, $\lambda_2^* = .24$, $\lambda_3^* = .40$, $\lambda_4^* = .32$

Therefore the compromised function is

$$.04c^1x-.24c^2x-.40c^3x-.32c^4x$$

The compromised function has been maximized over the same feasible region. The solution is as follows:

Profit	=	4,115 millions Turkish Lira
Manufacturing cost	=	1,110 millions Turkish Lira
Overtime in glazing	=	552 hours
Overtime in Assembly lines	=	33,543 hours

When we presented our compromised solution to the manager, he was satisfied with that and did not want any further solutions. We believe the decision maker would have asked for some other solutions if he had not had experience with the other two methods.

When we compare the optimal solutions (from the manager's point of view)obtained with these three methods, we see that all of them are fairly close to each other.

6. SUMMARY and DISCUSSION

The Belenson-Kapur method is not referenced very often in multiple criteria literature. We believe the main reason for this is that the method apparently does not converge. We use a part of the method, mainly the part that generates weights to find an efficient solution. We have applied the modified version of the method to four different production planning problems. In some of them we used the method by itself, in some others we use it together with some other MCDM methods. Our initial solution has always been "very close" to the decision maker's final choice. We are well aware of the sample size. However, we prefer experimenting on real life projects rather than on randomly generated problems.

The ITM is more recent than the CCM. In addition it generates both efficient extreme as well as nonextreme points. Why should we use the CCM if we have the ITM? First, there are fairly well-known multicriteria methods that generate only efficient extreme points. How representative are these efficient extreme points of the totality of efficient solutions in actual problems?

In our experience it was much easier for our decision maker to compare efficient solutions generated by the CCM. The method is appealing, but heuristic in nature as Zionts (1987 p.28) points out.

Our decision maker had a lot of difficulty in selecting a solution generated by the ITM. In our view this part of ITM might be improved by using any of the discrete multicriteria methods.

When we experimented with the ITM, each solution took approximately 5-7 minutes on IBM-XT for a problem with 4 objective functions 96 variables 137 constraints. Most of the constraints were upperbounds and/or lower bounds. We thought it would be nice if one could obtain one solution; then dynamically see the impact of changing one objective function on others. In this respect the VIG(Visual Interactive Goal Programming) package developed by Korhonen (1987) seems to be promising. We are currently experimenting with it. It has its own limitations too. For example, currently it can only handle 96 variables with 100 constraints out of which 10 of them could be objective functions.

The author wish to thank Professor Ram. M.R. Kasuganti (Youngstown State University), Professor Ralph E. Steuer (University of Georgia), and Professor Stanley Zionts (State University of New York at Buffalo) for their constructive comments and suggestions.

REFERENCES

1. Belenson, S.M. & Kapur, K.C. 1973 "An algorithm for Solving Multicriterion Linear Programming Problems with Examples.", O.R. Quarterly., 24(1), 65-77

2. Cunningham, K. & Schrage L. 1987 VINO, The Scientific Press, Redwood City, California

3. Karpak, B. 1988 Multiple Criteria Decision Support Systems in Production Planning, A Micro Experience, Working Paper, Youngstown State University, Youngstown, Ohio, USA.

4. Karpak, B. 1978 Cok Amacli Optimizasyon Modellerinin Irdelenmesi Oyunlar kuramina Dayali Etkilesimli Bir Algoritma Onerisi ve Cok Amacli Bir Uretim Planlama Surecine uygulanmasi(in Turkish). Unpublished Thesis to qualify as an Associate professor,Istanbul University, faculty of Business Administration,Rumelihisariustu,Istanbul,Turkey

5. Karpak, B. and Kuruuzum, A. 1987 "Contracting Cone Method vs Modified Two Person Zero-Sum Game Approach: An Implementation in Production Planning" in Y.Sawaragi,K.Inoue and H.Nakayama, (Eds.),Toward Interactive and Intelligent Decision Support Systems,Springer Verlag,New York

6. Korhonen, P.J. 1987 VIG (Visual Interactive Goal Programming) User's Guide

7. Korhonen, P.J.,& Wallenius, J., 1987 A Pareto Race , Unpublished Paper, Helsinki School of Economics.

8. Steuer, R.E. 1986. Multiple Criteria Optimization: Theory Computation and Application. Wiley, New York

9. Steuer, R.E. 1977. "Vector Maximum Gradient Cone Constraction Techniques. Lecture Notes in Economics and mathematical Systems. in Zionts(ed) 155 Springer Verlag, New york.

10. Steuer, R.E. & Schuler, A.T. 1983 Operating Manual For ADBASE Multiple Objective Linear Programming Computer package, College of Business Administration, University of Georgia

11. Steuer, R.E. & Choo, E.U. 1983 "An Interactive Weighted Tchebycheff Procedure for Multiple Objective Programming ", Mathematical Programming. 26(1), 326-344

12. Zeleny, M. 1987 "Systems Approach to Multiple Criteria Decision Making" in Y.Sawaragi,K.Inoue and H.Nakayama, Eds.),Toward Interactive and Intelligent Decision Support Systems,Springer Verlag,New York

13. Zionts, S. 1987 Multiple Criteria Mathematical Programming: An Updated Overview and Several Approaches, in this publication.

CONFLICTING OBJECTIVES IN SOFTWARE SYSTEM DESIGN

A. Ulkucu
AT&T Bell Laboratories
480 Red Hill Road
Middletown, New Jersey 07748
U.S.A.

1. *INTRODUCTION*

The main goal of this panel discussion was to analyze conflicting objectives in the design of software implementing Multiple Critieria Decision Making (MCDM) methods. A synopsis of the proposed scope of the meeting, presented in Appendix 1, was distributed in advance to the participants of the NATO Advanced Study Institute. Two of the panelists, Professors Tuncay Saydam and Tawfik Jelassi, were invited to prepare presentations, respectively, on the structural model of a software design process and MCDM software development by small teams, respectively. A list of attendees is given in Appendix 2.

An edited version of the transcript of the discussion is provided in Appendix 3. Key points are summarized in the following.

2. *SUMMARY OF DISCUSSION*

Effective use of computers in multiple criteria decision making requires proper planning for the application software, which itself is an MCDM process. Analysis of the MCDM system development requires a good understanding of the software design process. This process starts with modeling the real-life decision-making problem. The model characteristics define the MCDM system requirements, which give rise to a software model. The structural model of a software design process also includes programming language, application environment and software design team.

NATO ASI Series, Vol. F56
Multiple Criteria Decision Making and Risk Analysis
Using Microcomputers. Edited by B. Karpak and S. Zionts

The language choice criteria include: (1) Wide availability and applicability, (2) Ease of modular construction, (3) Data abstraction, (4) Information hiding, (5) Free formatting, (6) Ease of debugging, (7) Ease of interactive use, and (8) Extent of development tools. Some popular programming languages, such as C and ADA, satisfy many of these criteria.

The application environment consists of two components: the computer system and the decision application environment. The computer system includes the hardware and the operating system. The former should be widely used and the latter well known and widely available. The decision application environment includes all possible users of the developed software, and we would like the product to have as many users as possible.

The size and the composition of the design team are very important. While the size is determined by the complexity, the composition should include the following: analysts (decision scientists in the case of MCDM), software engineers, programmers, and system users. All these members must work very closely with the software manager, who should be in charge.

Some of the objectives in the MCDM software design are:

(1) Most reliable implementation of MCDM algorithm

(2) Highest level of modularity

(3) Lowest level of granularity

(4) Highest level of interactiveness

(5) Highest level of distributiveness

(6) Shortest development time

(7) Least costly development

(8) Highest level of system support

(9) Least costly software maintenance

(10) Longest product life.

Several of these are in obvious conflict with each other, such as objectives 2 and 3, 4 and 7, 7 and 9. Therefore, not only is the software product a tool for MCDM, but also the software development process itself involves an MCDM.

The logical architecture of a Decision Support System (DSS) implementing MCDM methods can be partitioned into three main components: (1) The mathematical algorithm, or model, component, (2) The data component, and (3) The user interface, or dialog, component. The main objectives in the design of these components are as follows.

In many situations, the decision maker has an organizational structure, such as a hierarchical one consisting of top management, middle management, and so on. Top management, for example, might have a long-term, relatively stable preference structure, whereas lower management might change its preference structure more often. The model should reflect the user's preference structure. One approach is to have a model bank including several MCDM algorithms, so that the user can choose the best one. Other objectives are to minimize both the response time and the development effort. In addition, we might want to integrate the algorithms so that the user has a unified access to them.

Several design decisions regarding the data component have also to be made, such as detailed versus aggregated data, current or historical data, etc. Here again one objective, which conflicts with the others, is to minimize the response time.

The user interface is the most important component, because for the decision maker, the system is whatever he/she sees at the user interface. The variety among decision makers--different backgrounds, experience and computer expertise--presents challenges to the user interface design. Furthermore, since the user interface facilitates communication between the decision maker, his/her mental model, and the knowledge stored in the DSS, we should try to make the user interface component as independent as possible from the decision problem at hand.

The overriding concerns in the development of a DSS product are the development time, which we want to minimize, and the quality of the product, which we want to maximize. Since the DSS may be seen as a threat to the decision maker, who thinks it is going to replace him/her, we also want to maximize the user's involvement in the development. One approach is to use prototyping, which may minimize development time. (There was no agreement on the impact of prototyping. It was also argued that prototyping can maximize the quality at the expense of increased development time.) Another approach is to use the structured development life-cycle (SDLC) methodology, which involves requirement analysis, logical and functional design, coding, testing and maintenance stages. Although the quality objective can be achieved, the cost will be high. Moreover, users cannot participate in the design as they are not experts in the SDLC methodology. A hybrid approach, a combination of prototyping and SDLC, may provide a compromise.

One of the issues regarding the user's involvement is the education of the decision makers, especially the ones who are not open-minded and are skeptical of analytical techniques, distrusting the "black box", the MCDM model. Before getting involved in the MCDM system development, they have to be introduced to the interactive, computer-based decision aids which help achieve better and faster decisions. The DSS labs, such as the one at Indiana University, may motivate the managers from industry to voluntarily learn about the DSSs.

Using an approach allowing the user's involvement, and with proper training, the decision maker could be involved in the development process from the ground floor. He/she could be responsible for establishing specifications and making sure that the software captures the specifications. He/she could also be made responsible for providing the data.

Another important factor is the analyst's training. It's very difficult to eventually deal with the decision maker, if the analyst has not had any exposure to the difficulties of making a real decision. A recommended solution is co-op educational programs for MCDM, which would allow interactions between the students who are going to be the next generation of engineers of MCDM systems and the people who would be using them. Therefore, the main contribution of industry/university cooperation in increasing the effectiveness of multiple criteria decision support systems is to start the co-op programs lacking in MCDM.

APPENDIX 1

CONFLICTING OBJECTIVES IN SOFTWARE SYSTEM DESIGN

SCOPE OF THE PANEL DISCUSSION

Effective use of computers in multiple criteria decision making and risk analysis requires proper planning for the application software, which itself is a multicriteria decision-making process. In this round table discussion, we will analyze conflicting objectives in software system design and investigate approaches to formulating a decision-making model.

A particular scenario will include a researcher, say a Ph.D student or a post-doctoral researcher, who would like to demonstrate his/her algorithm using a microcomputer and also develop a decision support system product which is based on the algorithm and can be used by decision makers. Such a scenario should ideally require involvement of a multi-disciplinary team whose expertise includes systems engineering, software engineering and human factors as well as operations research and product planning.

If forming such a team is not practical, then the researcher might consider expanding his/her knowledge to other areas before attempting to develop the decision support system. Whether there is a multi-disciplinary product development team or a one-person team representing various areas of expertise, it is crucial that the product model be developed. This model may include several conflicting objectives, such as:

NATO ASI Series, Vol. F56
Multiple Criteria Decision Making and Risk Analysis
Using Microcomputers. Edited by B. Karpak and S. Zionts

(1) Shortest development time

(2) Most reliable implementation of the decision-making algorithm

(3) Maximum code reuse

(4) Best documentation

(5) Most efficient user interface

(6) Largest number of users (potential customers)

(7) Longest product life

(8) Least costly software maintenance

Note that the type of conflict these objectives present depends on several factors, in particular development environment. For example, if the researcher is not familiar with or has no access to reusable software modules which are applicable to the product, objectives 1 and 3 could be conflicting.

The following issues are presented to the discussants for comments at the round table:

(1) What is the appropriate level of detail in formulating the objectives in developing software for multiple criteria decision making?

(2) What are the conditions that make the proposed multi-disciplinary development process effective?

(3) What should be the role of industry/university cooperation in increasing the effectiveness of multiple criteria decision support system development?

Our intention is to make this round table a meeting where many discussants can contribute their experience, views or suggestions so that we investigate the issues from several perspectives. Assuming such an active participation, the round table discussion is expected to produce a report for inclusion in the conference proceedings. A minimal goal for such a report should be a set of guidelines for young researchers in developing software that can not only demonstrate their multiple criteria decision making methods, but also facilitate using these methods by other decision makers

APPENDIX 2

CONFLICTING OBJECTIVES IN SOFTWARE SYSTEM DESIGN

PARTICIPANTS OF THE PANEL DISCUSSION

M. Ahlatcioglu
Y. Aksoy
E. H. Arnett
J. C. N. Climaco
M. A. A. Cox
M. J. Crespo
A. Durso
L. R. Gardiner
P. Gomez-Vilda
K. J. Heidel
V. Hertogh
M. T. Jelassi
I. Kara
B. Karpak
R. Kasuganti
M. Koksalan
S. Kondakci
L. Lillich
S. Nummi
P. Oliveira
A. F. Ozok
D. Rios-Insua
V. Rodellar
T. Saydam
C. Schneeweiss
M. Sivri
A. Sobrinho
S. Southgate
M. Sucu
P. Teixeira
M. A. Tumeo
A. Ulkucu
V. R. R. Uppuluri

NATO ASI Series, Vol. F56
Multiple Criteria Decision Making and Risk Analysis
Using Microcomputers. Edited by B. Karpak and S. Zionts

APPENDIX 3

CONFLICTING OBJECTIVES IN SOFTWARE SYSTEM DESIGN

TRANSCRIPT OF THE PANEL DISCUSSION

Ulkucu: Good evening, ladies and gentlemen. I would like to thank you for your participation in this round table. First, we will define the scope and goal of this discussion. Then, participants, hopefully all of us, will contribute to the discussion by presenting our viewpoints and our comments on others'. Finally we will summarize the main points raised at this meeting.

The scope has been established as follows: Effective use of computers in multiple criteria decision-making and risk analysis requires proper planning for the application software, which itself is a multi-criteria decision-making process. In this round table discussion, we will analyze conflicting objectives in software design and investigate approaches in formulating the decision-making model. Following issues are to be addressed:

1. What is the appropriate level of detail in formulating the objectives in developing software for multiple-criteria decision making?

2. What are the conditions that make the proposed multi-disciplinary development process effective?

NATO ASI Series, Vol. F56
Multiple Criteria Decision Making and Risk Analysis
Using Microcomputers. Edited by B. Karpak and S. Zionts

3. What should be the role of industry/university cooperation in increasing the effectiveness of multiple criteria decision support systems development?

A minimum goal for the meeting is to identify a set of guidelines for young researchers in developing software that can not only demonstrate their multiple-criteria decision-making methods but also help other decision makers use these methods.

Let us start the discussion with the scenario which we will address. Suppose the researcher, say a PhD student or a post-doctoral researcher, which we will call the analyst for brevity, would like to demonstrate his or her MCDM method using a computer, possibly a micro computer. Suppose further that the analyst desires to develop a decision support system product that is based on the algorithm. The motivation behind this desire is clear: (1) The product could be sold to bring revenue to the researcher's institution. (2) The software which is developed with such a product-oriented approach would make it easier for other researchers to use the algorithm bringing fame to the researcher. (3) The analyst can more easily experiment with the algorithm and thus can extend his or her research faster; that is, the researcher becomes more productive. It seems that this product-oriented approach would help the analyst increase his or her effectiveness and improve the return of the effort spent on the algorithm. We first have to investigate the feasibility of such an approach. To do this we have to understand the software development process. May I now invite Dr. Saydam, who has volunteered to talk on this topic?

Saydam: Thank you, Mr. Chairman. My brief presentation will emphasize the following: I will first present key elements of the structural model of a software design process. I will then elaborate on certain important elements of the design process, namely, the desired software characteristics, key design issues which involve programming language, design teams, requirements and application environment. I will then discuss the objectives and conflicting objectives.

I'd like to define a few pertinent terms. The first is software. It's a representation of some real-life system or a model thereof, in terms of syntactic and semantic framework of a programming language to run on a computer system. The software that I have in mind here exceeds 10,000 lines of code. The second term is software design process. It is a collection of formal or heuristic software and system engineering procedures employed in accomplishing the software as defined above.

The third term is objectives. It is a set of well-defined rules, guidelines, criteria or desires that one would like to optimize while satisfying the systems software, manpower, language, and other pertinent requirements.

The fourth and final term is conflicting objectives--objectives of which the simultaneous optimization cannot be accomplished without certain compromising trade-offs.

I will now try to present the structural model of a software design process. This process starts by identifying the real-life decision-making system. This system is reduced to a model by a set of abstractions. Model characteristics further define the requirement analysis and planning, which gives rise to, or become a component of, a software model. Software model is the key element in this structure. The others are programming language, application environment, and the software design team. The objectives that one considers as both conflicting and non-conflicting profoundly influence the abstraction process, requirement analysis, and software model. The feedback loops act on this model as follows: The software design team influences the design process, as well as the requirement analysis. The final goal of the whole process is obviously an implementation of some decision process in software. The feedback loops emanating from this implementation will go to the requirement analysis, to the model, to the real-life system, and finally to the objectives. Obviously the closed circle on the implementation is the life cycle of the software.

I will now describe the desired software characteristics, which include: (1) Correctness, correct and reliable implementation of the decision-making, (2) Robustness, (3) Modularity, (4) Updatability, (5) Maintainability, (6) Interactiveness, (7) User-friendliness, (8) Life cycle, (9) Cost, (10) User interface, (11) Documentation, (12) Development time, and (13) Applicability.

I will now briefly discuss the key design issues. The first is a programming language. Some of the more popular programming languages that I feel appropriate for such tasks are PASCAL, MODULO 2, C, and ADA. The language choice issues are the following: (1) Wide availability and applicability, (2) Ease of modular construction, (3) Data abstractions, (4) Information hiding, (5) Provision of high-level data structures, (6) Free formatting, (7) Ease of debugging, (8) Ease of interactive use, and (9) Extent of library facilities.

The second important element of a design issue is the design team. The size and the composition of this team are very important. The size obviously is determined by the complexity of the task involved. But the composition should include the following: Decision scientists, software engineers, programmer analysts, and system users. All these members must work very closely with the software manager who should be in charge.

The third key design issue is the application environment, which consists of two components. The first is the computer system, which involves the hardware and the systems software including the operating system. Keeping in mind that the developed software should have wide applicability, one should target it to a widely used hardware as well as a well known and widely available operating system. The second application environment component is the decision application environment. We would like the developed tool to be used by as many decision makers in as many different environments as possible.

I will now consider the objectives issue. One can call these desired properties of the MCDM software. We would like that this software to: (1) be highly modular, (2) be highly interactive, (3) have short development cycle, (4) have long life cycle, (5) be easy and inexpensive to maintain, (6) incur low cost of development, (7) have low level of granularity--the amount of detail involved in the model, (8) have high level of system support, (9) be easy to update, (10) have high-level of adaptability, and (11) be distributive with increase in size--if a system cannot fit into a given environment, there are tools to distribute it among a variety of hardware environments.

Of these, several are in obvious conflict with each other. Number 1 and 7; 2 and 6; 3 and 7; 10, 5, and 6 are the most obvious. I'd like to stress that one cannot optimize all the objectives simultaneously without compromising others.

I'd like to conclude with the following remarks. The importance and inclusion of conflicting objectives into a software design process cannot be overemphasized. Not only is the finished software product a tool for multi-criteria decision-making for conflicting objectives, but also the process of developing the tool itself is an MCDM under the conflicting objectives. With ever-increasing software development costs, relative to hardware, the use of MCDM techniques for software development becomes even more relevant and urgent. Thank you.

Ulkucu: Thank you, Dr. Saydam, for the excellent overview on the software development process, clearly a process that requires a multi-disciplinary development team, which we will have to discuss later in this meeting. Let us now focus our attention on designing and building interactive decision support systems, which is not a trivial task. Various issues need to be analyzed. We need to address software design objectives in decision support systems and their conflicting nature in more detail. Now I would like to invite Dr. Jelassi, who has volunteered to talk about these issues.

Jelassi: Thank you. The previous intervention dealt with large scale systems and a development team instead of a single person in charge of building the system. In my presentation, I will cover the other extreme of the spectrum. Suppose you are a consultant or an information systems professional. You have been charged by your manager, or you have been called by a firm to build an interactive decision-aid system. The questions that need to be answered are: what are the conflicting objectives in software development, what are the decisions that need to be made in system design, and how to go about making these decisions.

First of all, let me make two assumptions: (1) This one-person team is skilled in building computer-based information systems, and (2) the product that we want to deliver is a decision support system (DSS). The approach that will be taken is based on Sprague and Carlson's DSS framework that consists of three system component. The first is a model component that consists of the mathematical algorithm to be executed against a set of data stored in the system. The second component is either a database management system or a traditional file management system, which handles the stored data. Obviously, to allow a decision maker or a staff member who might assist the ultimate decision maker in this process, we need to have a user interface (or dialogue module)--that is, the third system component.

Now I would like to mention the conflicting design objectives in each of these three system components. First, let us start with the model module. Since our one-person team is made of an information system expert, for him or her the mathematical algorithm is a black box. He or she doesn't necessarily have the sufficient background in operations research or management science to be able to understand the decision model. This developer, however, is interested in knowing two things. First, what is the set of inputs that this black box is going to process? In the case of MCDM, the inputs are the set of criteria, possible alternatives, the weights to assign to these criteria, and maybe additional parameters and coefficients. Then this data will be executed against the mathematical algorithm of the MCDM model to deliver some outputs, which could be in a form of ranked alternatives, utility values associated with them, optimal solution, etc.

What are the objectives, what are the conflicts here?

Objective 1. Have in the DSS not just one MCDM method, but two or more methods, so we can enrich our system. If the user wants to use additive utility functions, the system has them stored. If he/she prefers to use multiplicative utility functions or other methods, the system should be able to support that. So we'd like to have a "model bank", a set of methods and not a stand-alone algorithm.

Objective 2. Link these methods in a way that allows the outputs of one method to be used as the inputs to another. Doing so provides some integration among these methods.

Objective 3. Minimize the DSS response time to avoid potential frustration and impatience on the part of the decision maker. Doing so implies reducing data retrieval time as well as the execution time of the mathematical algorithm.

Objective 4. Minimize the time of initial development, as well as the time of database update, model bank restructuring, and maintenance. Here, clearly we have a conflict. For example, the model base management system (which supports method definition and restructing) will increase the execution time in the system and consequently delay DSS responses to user requests.

Let me now shift to the second component of the system, the data module. The data module is becoming the cornerstone of interactive decision-aid systems, due to the increasing volume of data to process, and to the downloading/uploading operations needed in micro-mainframe computer configurations. Moreover, a variety of data is stored in the system, including criteria, alternatives, intermediate as well as final results. We have objectives and conflicts here as well:

1. Provide data in both detailed and aggregated form.

2. Provide both current and historical data.

3. Support access to both internal (departmental or corporate) and external (e.g., purchased from a market research institute) data.

4. Support access to both private and public data.

5. Support the decision maker in accessing the data on an ad-hoc basis, without having to have his requests pre-defined in the system.

6. Provide the decision maker with simple modeling operations through the use of the database management system (specifically by executing built-in functions, such as min, max, mean, standard deviation, etc.).

7. Guarantee a quick response time.

An example of conflict here is between the seventh objective and the first six. One may argue that commercially available database management systems satisfy the requirements (objectives) listed above. In fact that is not true. If you look at any database management system available in the market, you will not find one that supports all the requirements for MCDM methods.

Let me now focus on the third component, the user interface. This is the most important module of the system, because, for the decision maker the DSS is what he sees displayed on the screen. The objectives and conflicts here are as follows:

1. Support a variety of decision makers--different backgrounds, computer experiences, and cognitive styles. To achieve this objective, the dialogue component should contain several interaction modes: question/answer dialogue, menu selection, command language (or type-ahead mode), graphic representation (such as icons, maps, etc.), semi-natural language, etc. This variety of interaction styles facilitates users' dialogue interaction preferences and copes, over time, with their changing cognitive styles.

2. The dialogue component is the communication medium between the decision maker and his implicit mental model on one hand, and the knowledge built in the DSS on the other hand. The objective here is to make the dialogue module as independent as possible from the decision problem at hand, in order to meet the expectations of the user's mental model.

3. The third objective deals with the issue of user-friendliness, specifically, minimizing the training time, re-learning effort, the types and rates of usage errors, and the difficulty of recovering from these errors.

The objectives listed above conflict with the desire to have a quick response time. Some empirical studies have shown that to make a DSS product acceptable and usable by decision makers, the system response time must be less than two seconds. You can imagine the technical challenge that the DSS developer faces!

So far I have mentioned the different objectives and the conflicts for the three DSS components: the model, the data and the dialogue modules. Let me now conclude with some overall objectives that the system must also satisfy:

1. Minimizing the development time and cost.

2. Maximizing the quality of the delivered system. We don't want to have a quick and dirty solution. Quality, acceptance and usage of the DSS strongly depend on the level of user involvement in the development stage. This latter factor can also help in avoiding (or at least reducing) the decision maker's resistance to the system that he/she might view as a threat to his/her decision-making job.

The challenge here is how to meet these system development objectives. One possible approach is to use a prototyping approach; that is, to first build a very small system (say version 0) and show it to the manager. The next step is to expand on it (to come up with version 1) and seek user feedback. It's an iterative process, relatively quick and cheap.

The prototyping approach minimizes the development time and cost, while maximizing user participation. It contrasts with the structured development life-cycle (SDLC) methodology, which consists of the following steps: requirements analysis, feasibility study, logical design, physical design, programming, testing, implementation, and maintenance of the system. Compared to prototyping, SDLC is an expensive and time-consuming approach.

Here are some thoughts that I've had for the last couple of days, since I was asked to participate in this panel. I wanted to share them with you and hopefully stimulate the debate among us. I would like to conclude this presentation by saying that software design is a multiple criteria decision-making problem by itself. There are several objectives which are conflicting by nature; the system designer cannot attain one objective unless he does it at the expense of another. In most cases, trade-offs are necessary to compromise between extreme choices. The quality of the end product in general (or the DSS here) does strongly depend on those decisions that we make at the design stage of the system development process. Thank you for your attention.

Ulkucu: Thank you, Dr. Jelassi for this very interesting overview. I believe it is now time to open the floor to all the panelists for suggestions and comments. One of the things that I would like to see done during this panel is a list of objectives on which we will all agree, along with a priorization of the list. We already have three lists of objectives: the one presented in the statement of scope of the panel, which I distributed two days ago, and the ones given in Dr. Jelassi's and Dr. Saydam's presentations. These lists mostly agree. We need a discussion on common objectives, at least the important ones.

Tumeo: My name is Mark Tumeo. Thinking over the things brought up by both of the presentations, I suggest that at the top of the list of objectives should be including the user in the process. In his presentation Professor Saydam said that the software manager is the person who should be in charge. As a consultant hired to build software for various decision-making activities, I don't believe that. In actuality that's not the way it works; it's the users who are in charge. They are the ones paying for development, and they are the ones using it. The users should be in on the ground floor, working on questions like: What's the model going to do? What is the purpose of the model? What kind of things do you want it to do? And, if there is any kind of iteration, it should be with that user. Bring the version to him or her, and have him or her say, "Well, you know I don't particularly like the way it formats the output, can I change this for this month?" Let the users ask various questions, what-if type of questions, to make sure that the model fits their needs and that they are very clear about what the model is doing. I think that the user interaction in the development phase is the most important thing in developing software.

Durso: My name is Tony Durso. I agree with Mark (Tumeo). I'd like to take it one step further. Would Professor Jelassi put the first slide up? I think that the comment about the user being ultimately responsible implies that he is in charge. This room is filled with analysts. I have had the opportunity of being an analyst, a decision maker, and what you call a dialogue component. I've done all three. The system I have in mind was exceptionally large and it required the coordination of multiple agencies, which made it even more complicated. Now in your case, you are talking about a private consulting firm of some sort that supports another agency. I would say for that situation there is a set of rules, as opposed to, let's say, an internal system development agency. Different things need to be done. But in any case, whether it's an internal consulting agency, or an external consulting agency, as it is shown here, the decision maker's being in charge through his dialogue component is absolutely critical. The decision maker should be involved from the ground floor. He is responsible for establishing specifications. He is responsible for meeting with the analysts who are actually writing the software, to make sure that it captures the specifications, that they are not misinterpreted, and that they actually work. He is also responsible for providing the data. The data is also quite difficult to get, because it may not be exactly in the form you need to run your model. So, the decision maker is really the area expert, and therefore is even more critical than has been suggested. I'd solicit anybody else to add to that.

Ulkucu: Do we have any comments?

Schneeweiss: I'm Christoph Schneeweiss, from the University of Mannheim. In many cases the decision maker is not only one person, but a whole group of persons, which has a certain structure. There is for instance top management, middle management, and so on. So, in many situations we have to construct multi-criteria decision models which reflect the hierarchical structure of an organization. That is, it might well be necessary to construct a model with parameters which have to be fixed by top management whereas other parameters have to be set by lower management. Often there's a certain preference structure which is given by top management in a very aggregate way, and on the other hand there's a certain preference structure which is given within this framework by lower management. That is, top management might fix a long term, comparatively stable preference structure whereas the lower management levels might be allowed to change their preference structure more often. So I'd like to see this structure also in our multi-criteria decision models. It might well be that by exploiting the hierarchical structure of an organization, some of our problems of modeling the preference information for a real world application would vanish, simply because "hard facts" given by the upper management levels restrict the freedom one has in modeling the preference information on the lower level. Thank you.

Ulkucu: Thank you, Dr. Schneeweiss. Any more comments?

Heidel: Mr. Chairman, my name is Karen Heidel. I'd like to first compliment Dr. Saydam and Dr. Jelassi for their sense of vision. I do believe, as an applications individual who takes the things that are developed and applies them in the real world, that a sense of vision is important. I believe that if we don't know where we want to be in the future, it will be very difficult to get there in a random process. These two gentlemen have a very good feel for the requirements for using MCDM in real world applications. Primarily, in terms of the development of software, there are marketing issues to be addressed. Very real-world kinds of marketing issues exist in terms of increasing the acceptance of MCDM. It is very difficult to walk up to an individual who is highly placed and powerful, and say to them, "You don't know how to make this decision." Yet, that is the kind of difficulty now being experienced. So as we envision a world where everybody has a PC, which is interactive, and everybody has a variety of MCDM models from which to chose, we need a sense of vision and direction, one that will increase the utility of the multiple criteria solution processes well into the future.

Ulkucu: Thank you, Ms. Heidel. Are there any more comments?

Saydam: I would like to comment on the user involvement in the software design process. I'd like to differentiate between a user involvement and user becoming a software manager. I believe that users should be involved in every phase of this development process. But managing a software team is an established discipline that the users, more often than not, are not prepared to undertake.

Therefore, my remarks were quite pertinent, with respect to having a software engineer manage a highly technical software process, while having a great degree of user involvement in almost every phase of this process. I totally agree with my friends that it is the users who bring in the requirements, the specification and the data, but I wouldn't go as far as having a user preside as a software manager.

Ulkucu: Thank you. This opens the discussion to another area. The recent advances in user interface technology, and software development in general, make it possible for users to design user interfaces without actual programming. Therefore, software engineering should be included in the process so that user involvement can be possible through prototyping, using the tools provided by the customizable, flexible interfaces.

Gardiner: I'd like a clarification in Professor Jelassi's description of overall objectives. When you mention that prototyping may not maximize software quality, I wonder if there might be times when prototyping would maximize software quality. It would seem to me that the time lag in the traditional systems development life-cycle might seriously impair software quality. And also there might be the type of difficulties that occur when users don't use the screens and discover, after two years of development, that the information they thought would be there isn't there. I just wanted a clarification on that.

Jelassi: In my opinion, prototyping is not the most appropriate approach to use if we want to maximize the quality of the end product. Regarding end-user computing, there are some traps that should be avoided. The quite novel idea of letting users develop their own computer-based information systems is an attractive one. Recent technological developments (such as fourth generation languages, expert systems, artificial intelligence, DSS generators, and expert systems shells) can make it somehow possible. However, the average user does not have sufficient technical skills to undertake the software building task alone. Having said this, I agree with you concerning the need to involve, as much as possible, the user in the system development process, especially if we want to deliver a DSS that will be acceptable and usable by the decision maker. To avoid the traps I mentioned earlier, having a user develop his/her system with the assistance of an information system specialist could be seen as a practical compromise and a good combination of functional and technical skills.

Ulkucu: Thank you, Dr. Jelassi. As a person who is working on user interfaces and trying also to apply OR techniques to user interface development, I'm very happy to see the importance attached to user interfaces. Unfortunately, we don't have time in this panel to expand on them. We should postpone doing so to another occasion, perhaps a specific meeting on user interfaces. But I would like to make one comment, however, that software quality may or may not contradict prototyping, depending on the conditions. For example, there are software development tools that greatly facilitate prototyping. If these tools are used, prototyping can actually increase quality. Under such conditions, the MCDM analyst should consider using prototyping. In other cases, he/she may choose to avoid it. Therefore, the system development objectives may or may not conflict depending on the development conditions, which have to be identified at the beginning of the process.

Now, let me reiterate one of our goals, and I would like to ask your comments. Is there anything that university/industry cooperation can do to increase the overall effectiveness of the efforts on MCDM in academia as well as industry? Can we take something from this meeting, some simple guidelines, to guide future projects?

Tumeo: I suppose this is rather mundane. I am just coming out of a PhD program and starting as a professor in a university. One of the things that I think is necessary, especially in my branch of engineering, civil and the use of computers there, is cooperation. Universities love to do research, I love to do research projects, I can always say that industry should find more research for the university and vice versa. One thing that would be very helpful in this field is to have some interactions between the students who are going to be the next generation of engineers and people using MCDM methods, getting hands-on experience in industry, some kind of co-op programs. I don't know of any co-op programs in MCDM or the use of computers for MCDM. As a consequence, here I am finishing my PhD, and, in my field of engineering, this conference is actually the first exposure I have had to the technical language associated with

MCDM. I'd say that would be a lack in my education. And as a professor or practicing engineer, that lack is going to cause me some trouble in using the MCDM software in the future. I think that if you're going to make a big impact, in the long run, you need to start at the education level and get some cooperative type programs for the students and industry going.

Durso: I'd like to return to one point I made before, and expand upon Mark's comment. It's very difficult to be an analyst who must eventually deal with the decision maker, if the young analyst has not had any exposure to the difficulties of making a real decision in industry, in government, or wherever. In that cooperative program, there should be an opportunity to be exposed to a decision maker making real decisions, and see what he has to go through to finally come down to the ultimate decision which is going to affect the success or failure of his organization. That may be going one step too far. But you can certainly appreciate the value of your product if you are on the receiving end and you are put in a precarious situation of having to make a life-and-death decision with a limited support system. That kind of exposure is probably asking a lot, but I think that's the ultimate requirement.

Ulkucu: Thank you, Mr. Durso. May I simplify or reinterpret the guideline you suggested--that the young analyst should try to gain experience in decision-making?

Durso: Yes, absolutely.

Ulkucu: And may I add one more? The analyst should leave room for others in his or her project. For example, try to attract a human factors person to help you design the user interface. We all understand now how important it is. I believe a PhD project in MCDM can very well nourish, for example, at least an MS project in human factors. So don't try to do it all, try to form a team.

Durso: I'd like to add a little bit more information and leave the university. I know the point I'd like to add is not in the university environment as far as students are concerned. But I would like to share some information or perceptions of decision makers about analysts. We have probably touched upon it at one time or another during the course of the institute. I have had the opportunity to work at very high levels and have had the good fortune to work for open-minded as well as not so open-minded decision makers. The ones who were open-minded to analytical techniques are not difficult to deal with. Generally they are pretty bright, and they like the work that we do. The problem is dealing with the one who is not open-minded. He is very skeptical of analytical techniques, he distrusts the black box, he distrusts the analyst because he tends to work in a vacuum. And that's why I made the point before about the decision maker being actively involved in the system development. It is of the utmost importance that the decision maker be assured that the analyst is not working in a vacuum, and that he is not on another plane (as far as thinking processes are concerned), that he has common sense, that he is reliable and is dedicated to getting the project done, and done correctly.

It's the analyst's image that probably causes him more consternation than the quality of his work. Unfortunately, I think, a lot of the high-level analysts tend to get stereotyped by high-level decision makers. The only way to fix that is for the analyst to work openly with decision makers, not get painted into a corner, or work in a vacuum.

Ulkucu: Thank you, Mr. Durso. I think this is again another excellent explanation of the thesis that the analyst should treat the implementation effort of the algorithm as a real product. Therefore, the analyst should expend effort to understand the user needs. This product-oriented approach will make the MCDM software development effort more successful. Do we have some comments on this issue, perhaps from graduate students?

Kasuganti: What I hear so far about user-friendliness is very well documented in the literature. The DSS field is looking into it now. However, regarding the comment that was made a little bit earlier, "What if the decision maker is not open-minded?", what can anyone do? Not much! If he refuses to do anything, we go around in circles. We should definitely try to develop user-friendly software; there is no contention on that. However, from the discussion I have heard so far, we are discussing mostly the people who don't want to use anything like these methods. There is not much anybody can do about it.

Jelassi: I think it is our responsibility as educators, researchers, and information systems developers to increase the level of awareness of managers concerning the potential contribution of DSS to the decision-making activities of the organization. Let me use an example here. At Indiana University, one research direction deals with the development of Group DSS (GDSS) and the assessment of the impact of this technology on decision-making tasks, behavior, performances, etc. We send copies of our GDSS research reports to the Department's corporate affiliates and invite decision makers to visit us and try the facility. If they cannot come, we let them know that our GDSS lab is portable, so we can take it to them. Other educational means include organizing conferences, publishing in journals and magazines. I believe these means should help in solving (or at least alleviating) the lack of awareness problem.

Kasuganti: That itself suggests that these people, who come to your lab or want your lab to go to them, are open to your suggestions. We are not in contention there. For those who refuse to do it, though, there is not much you can do. That's what I meant.

Ozok: Ahmet Ozok from the Istanbul Technical University. This is not a comment but a clarification from Professor Saydam. As a human factors experimenter, I am very interested in user-friendliness. Does user-friendliness include all the factors that we understand from the human factors discipline?

Saydam: Yes, I believe so. It involves factors like human psychology, reaction to response times, the ease with which the user interacts with the screen--it includes all these things. While I have the floor, may I shift the emphasis from user-friendliness to another topic which I think is important--one should not see MCDM software as a panacea for decision making. It may be a good tool, but we should not forget that the black box is acting as a model for decision-making. Therefore, the assumptions, the algorithms, the techniques that the black box has absorbed determine how far or near the whole software represents a real decision-making environment. Therefore, it should be clear to the user what is in the black box and what kinds of restrictions the whole software might have. I do believe that an MCDM software is an excellent tool for a variety of users. Now let me elaborate on some of its not so often mentioned uses. I think it might be an excellent training tool for decision makers. Usually, when a person makes the decision, one may not know how the variety of parameters might bear on the final decision. By using such a tool, one can tune and change these weights, the parameters, and do sensitivity analysis to see how the decisions change. It can be used as a simulation tool to analyze unusual decision making alternatives. In other words, one can manipulate the input structure of the whole system and see what sorts of decisions or set of decisions might come out of the software. The tool can be used to understand the decision making process itself. So I would like to reiterate the fact that, the way I view the MCDM is not only as a help for a manager to make a one-shot decision, but for a variety of other uses. After all a large MCDM package might be a very costly affair. One cannot afford to use it for making only a single decision.

Ulkucu: These comments again underline the fact that MCDM-type OR methods, and the software implementation of them, are quite different from some classical OR algorithms, because in the MDCM case, probably there are many users: decision makers, analysts, colleagues, etc. So, I think that underlines the fact that we should take very seriously the development of the software and look at all the aspects. But it is also clear that the development may be very costly. However, I have already noted several suggestions on how to lower the cost, for instance, using somebody else's code because of modularity. As was also discussed earlier during this institute, it is possible to come up with a decent code which is reusable by others if we follow certain rules.

Are there any comments from those who haven't spoken yet? I would like to conclude after two more comments.

Jelassi: I would like to make some concluding remarks about the decision-making environment. We are talking about building interactive, computer-based decision aids to help managers make better and faster decisions. These decisions are mainly tactical or strategic, and semi- or unstructured in nature. What also should be taken into account here is the level of interdependency of the decision task at hand--whether it is independent (and here personal support is required), or pooled (and here group support is needed), or sequential (which implies providing organizational support). I believe this last dimension, in addition to managerial level and structure of the task, should be integrated in the DSS design framework, if we want to develop a system that meets not only the technical requirements but also the organizational and behavioral ones, and consequently has a high chance for success. Thank you.

Ulkucu: Thank you, Dr. Jelassi. I suggest we conclude the panel with a review of the most important items. What is the most important thing that we have identified?

Southgate: Any system that you design must work for the user. It sounds silly, but it is possible to come up with something that looks very elegant, and satisfies you, but doesn't work. Unfortunately, generally speaking, you are not aiming at satisfying yourself. You're supposed to do a task for the user. You've got to assess his needs. I'd also take a slight exception to what Tawfik (Jelassi) said earlier about prototyping not maximizing quality. To me, prototyping has only one aim, and that is to maximize the quality of the product: not to speed up development particularly, but to make sure that the user is involved from the earliest possible stages, so that in his eyes the quality of the product to do his task is maximized. The majority of users I've come across when designing systems for them, have often very little idea of what they want. As you further refine their requirements, they often begin to realize what they want. When you have a user who has not really formulated in his own mind what he is after, I believe, prototyping can sometimes maximize development time. So I would take a slight exception to what you said there, although you are basing your comments on experience. I think it's dangerous to talk in black and white terms. You can't, especially when human beings and systems are concerned. Because things are not black and white, they are always grey. To me that's the most important thing. The system must do what the user wants. The person who writes the system, who designs it, is irrelevant. You're not there to satisfy your own needs, but purely the user's.

Ulkucu: Thank you Mr. Southgate. I knew that we'd have one more comment.

Durso: You have another one. Back to the element of time. Steve (Southgate) just indicated how it's more important to be correct than to be speedy. This is generally done under the auspices of a contract, which implies time. Get the product done in the time given, case closed! As a result, time may restrict an analyst from doing quality work.

Ulkucu: I would like to thank all of you for this very interesting panel discussion. Good night.

NATO ADVANCED STUDY INSTITUTE PARTICIPANTS

Multiple Criteria Decision Making and Risk Analysis
Using Microcomputers
June 28-July 8,1987 Tarabya Hotel,Tarabya, Turkey

Ahlatcioglu,M. (Dr.)
Yildiz Universitesi
Fen-Edebiyat Fakultesi
Sisli,ISTANBUL
TURKEY

Aksoy, Y. (Ms.)
University of Florida
BUS 219
Gainesville, FL 32611
USA

Ahmet, N.A. (Mr)
P.O. Box 1240
Ministry of Higher Education
and Scientific Research
BAGHDAD IRAQ

Assimakopoulos, N. (Dr.)
Organization Division
Agricultural Bank of Greece
173, Sigrou street
GR-171 21 ATHENS GREECE

Brans, J.P. (Prof.)
Universite Libre de Bruxelles
Pleinlaan 2
B-1050, BRUSSELS
BELGIUM

Arnett, E.H. (Mr.)
Dept. of Engineering and Pub.Pl
Carnegie - Mellon University
Pitsburg, PA 15213
USA

Bayazit, N. (Prof.)
Mimari Tasarim Yontemleri
Kursusu,
Teknik Universite
Taksim, ISTANBUL, TURKEY

Barda, H. (Mr.)
LAMSADE Universite Paris Dauphin
PL.Du Marechal de Lattre de
Tassigny,
75775 Cedex 16 Paris FRANCE

Bana e Costa, C. (Eng.)
CESUR
Instituto Superior Tecnico
Avenida Rovisco Pais
1000 LISBOA PORTUGAL

Cox, M.A.A. (Dr.)
Dept of Engineering Mathematics
University of Newcastle
Newcastle upon Tyne
NE1 7RU UK

Durso, A. (Mr.)
8013 Sunset Path Court
Springfield
Virginia 22153 USA

Climaco, J. (Prof.)
Universidade de Ciencias e Tech.
Departmento de Engenharia
Electrotecnica
3000 COIMBRA
PORTUGAL

NATO ASI Series, Vol. F56
Multiple Criteria Decision Making and Risk Analysis
Using Microcomputers. Edited by B. Karpak and S. Zionts

Dyer, J. (Prof.)
Department of Management,
CBA 4.202
University of Texas,
Austin, TX 78712 USA

Crespo, M.J.P.D.F (Ms.)
Rua Antonio Jose d'Almeida 12 6C
2780 OEIRAS
PORTUGAL

Erenguc, S.S. (Prof.)
Dept. of Management
University of Florida
219 Business Building
Gainesville, FL 32611 USA

Evren, R. (Prof.)
I.T.U. Isletme Fakultesi
Macka, ISTANBUL
TURKEY

Fandel, G. (Prof.)
Fernuniversitat Hagen
Feithstr 140
D- 5800 HAGEN 1
FRG

Isermann, H. (Prof.)
Fak.fur Wirtschaftswissenschaft.
Universitat Bielefeld
Postfach 8640
D-4800 BIELEFELD 1 FRG

Grauer, M. (Prof.)
FernUniversitat Gesamthochschule
Postfach 9 40
D-5800 Hagen
FRG

Gardiner, L.R. (Ms.)
M308 Rogers Road
Athens, Georgia 30605
U.S.A.

Gomez-Vilda,P. (Prof.)
Dept. of Electronica
Facultad de Informatica
CTRA de Valencia, KM. 7,00
28031 MADRID
SPAIN

Giresunlu, I.M. (Mr.)
Saithalimpasa cd. 93/1
Yenikoy, ISTANBUL
TURKEY

Gencyilmaz, G. (Prof.)
Faculty of Business Adm.
Istanbul University
Rumelihisariustu, ISTANBUL
TURKEY

Iz, P. (Prof.)
Department of Decision Sciences
Miami University
Oxford, OH 45056
USA

Heidel, K.J. (Ms.)
5741 East Paseo Cimarron
Tucson, AZ 85715
U.S.A.

Kara, I. (Prof.)
Anadolu University
Dept. of Industrial Eng.
ESKISEHIR TURKEY

Jelassi, M.T. (Prof.)
M.I.S. School of Business
Indiana University
10 th & Fee lane
Bloomington, IN 47405 USA

Hertogh, V. (Ms.)
Weesdijk 31
4132 BK VIANEN
HOLLAND

Kasuganti, R. (Prof.)
Dept. of Management
Youngstown State University
Youngtown, OH 44555
USA

Karpak, B. (Prof.)
Dept. of Management
Youngstown State University
Youngstown, OH 44555
USA

Kavrakoglu, I. (Prof.)
Bogazici University
Bebek
ISTANBUL TURKEY

Koksalan, M. (Prof.)
Industrial Engineering Dept.
Middle East Technical Univ.
ANKARA TURKEY

Kondakci, S.(Prof.)
Industrial Engineering Dept.
Middle East Technical Univ.
ANKARA
TURKEY

Kalipsiz, O. (Ms.)
I.U. Meslek Yuksek Okulu
Universite, ISTANBUL
TURKEY

Lillich, (Mr.)
Lehrstuhl fur Betriebswirtschaft
und Unternehmensforrschung
Universitat Mannheim
Schlos, D-6800 MANNHEIM 1
FRG

Orhun E. (Dr.)
Bilgisayar Biliml.Muhendisligi
Ege Universitesi
Bornova IZMIR
TURKEY

Kuruuzum, A. (Dr.)
Yildiz Universitesi
Fen-Edebiyat Fakultesi
Sisli, ISTANBUL
TURKEY

Pirlot, M. (Mr.)
Universite Libre de Bruxelles
CP 210, Boulevard du Triomphe
B-1050, BRUXELLES
BELGIUM

Nummi, S.(Ms.)
Helsinki School of Economics
Runeberginkatu 14-16
00100 HELSINKI
FINLAND

Mourmouris, J. (Dr.)
O.R. Management
25, Ir. Moutsopoulou Str.
18540 Passalimani
PIREOUS GREECE

Oliveira, P. (Prof.)
Pavilhoes de Engenharia
Universidade do Minho
4700 BRAGA
PORTUGAL

Sobrinho, A. (Prof.)
R. Do Jornal de Noticias, 15, 3
4100 PORTO
PORTUGAL

Ozkaya, G. (Prof.)
Faculty of Business Adm.
Istanbul Technical University
Macka, ISTANBUL
TURKEY

Ozok, A. (Prof.)
Faculty of Business Adm.
Istanbul Technical University
Macka, ISTANBUL
TURKEY

Raynaud, H. (Dr.)
IMAG LSD
BP 68
38402 St-Martin-d'Heres cedex
FRANCE

Saydam, T. (Prof.)
Department of Computer &
Information Sciences
University of Delaware
Newark, Delaware 19716
U.S.A.

Rios, D. (Mr)
Isaac Peral, 5
28015 MADRID
SPAIN

Simga, C. (Dr.)
Faculty of Business Adm.
Istanbul University
Rumelihisariustu
80830 ISTANBUL TURKEY

Serarslan, N. (Prof.)
ITU Isletme Fakultesi
Macka, ISTANBUL
TURKEY

Sivri, M. (Dr.)
Faculty of Science
Mimar Sinan University
Besiktas, ISTANBUL
TURKEY

Tumeo, M. (Mr.)
Department of Civil Eng.
University of California, Davis
DAVIS, CA 95616
USA

Schneeweis, C. (Prof.)
Lehrstuhl fur Betriebswirtschaft
und Unternehmensforrschung
Universitat Mannheim
Schlos, D-6800 MANNHEIM 1 FRG

Spronk, J. (Prof.)
Faculteit der Economische
Wetenschappen
Erasmus Universiteit Rotterdam
Postbus 1738 3000 DR ROTTERDAM
The NETHERLANDS

Steuer, R. (Prof.)
Dept. of Management Sciences
The University of Georgia
Brooks Hall
Athens, GEORGIA 30602
USA

Sucu, M. (Dr.)
Hacettepe University
Faculty of Science
Department of Statistics
Beytepe, ANKARA
TURKEY

Rodellar V. (Dr.)
Dept. Electronica
Facultat de Informatica
Ctra de Valencia, KM. 7,00
28031 MADRID
SPAIN

Teixeria, P.C. (Ms.)
Div. Quimica
Dept Geociencias
Quinta dos Prados- UTAD
5000 VILA REAL
PORTUGAL

TANYAS, M. (Prof.)
I.T.U. Isletme Fakultesi
Macka
ISTANBUL
TURKEY

Tversky, A. (Prof.)
Department of Psychology
Stanford University
Palo Alto, CA 94305
USA

Uppuluri, V.R.R. (Dr.)
Oak Ridge National Laboratories
P.O. Box Y, Building 9207A, MS2
Oak Ridge, Tennessee 37831
USA

Vincke, Ph. (Prof.)
U.L.B., C.P. 210
Boulevard de Triomphe
B-1050 BRUSSELS
BELGIUM

Wallenius, J. (Prof.)
Department of Economics
University of JYVASKLA
Seminaarinkatu 15
SF-40100 JYVASKLA 10
FINLAND

Yu, P.L. (Prof.)
School of Business
University of Kansas
Lawrence, KS 66045
USA

Ulkucu, A. (Dr.)
AT & T Bell Laboratories
480 Red Hill Rd.
Middletown, N.J. 07748
USA

Zionts, S. (Prof.)
School of Management
State University of New York
at Buffalo
Buffalo, NY 14260
USA

NATO ASI Series F

Including Special Programme on Sensory Systems for Robotic Control (ROB)

Vol. 1: Issues in Acoustic Signal – Image Processing and Recognition. Edited by C. H. Chen. VIII, 333 pages. 1983.

Vol. 2: Image Sequence Processing and Dynamic Scene Analysis. Edited by T. S. Huang. IX, 749 pages. 1983.

Vol. 3: Electronic Systems Effectiveness and Life Cycle Costing. Edited by J. K. Skwirzynski. XVII, 732 pages. 1983.

Vol. 4: Pictorial Data Analysis. Edited by R. M. Haralick. VIII, 468 pages. 1983.

Vol. 5: International Calibration Study of Traffic Conflict Techniques. Edited by E. Asmussen. VII, 229 pages. 1984.

Vol. 6: Information Technology and the Computer Network. Edited by K. G. Beauchamp. VIII, 271 pages. 1984.

Vol. 7: High-Speed Computation. Edited by J. S. Kowalik. IX, 441 pages. 1984.

Vol. 8: Program Transformation and Programming Environments. Report on a Workshop directed by F. L. Bauer and H. Remus. Edited by P. Pepper. XIV, 378 pages. 1984.

Vol. 9: Computer Aided Analysis and Optimization of Mechanical System Dynamics. Edited by E. J. Haug. XXII, 700 pages. 1984.

Vol. 10: Simulation and Model-Based Methodologies: An Integrative View. Edited by T. I. Ören, B. P. Zeigler, M. S. Elzas. XIII, 651 pages. 1984.

Vol. 11: Robotics and Artificial Intelligence. Edited by M. Brady, L. A. Gerhardt, H. F. Davidson. XVII, 693 pages. 1984.

Vol. 12: Combinatorial Algorithms on Words. Edited by A. Apostolico, Z. Galil. VIII, 361 pages. 1985.

Vol. 13: Logics and Models of Concurrent Systems. Edited by K. R. Apt. VIII, 498 pages. 1985.

Vol. 14: Control Flow and Data Flow: Concepts of Distributed Programming. Edited by M. Broy. VIII, 525 pages. 1985.

Vol. 15: Computational Mathematical Programming. Edited by K. Schittkowski. VIII, 451 pages. 1985.

Vol. 16: New Systems and Architectures for Automatic Speech Recognition and Synthesis. Edited by R. De Mori, C.Y. Suen. XIII, 630 pages. 1985.

Vol. 17: Fundamental Algorithms fcr Computer Graphics. Edited by R.A. Earnshaw. XVI, 1042 pages. 1985.

Vol. 18: Computer Architectures for Spatially Distributed Data. Edited by H. Freeman and G.G. Pieroni. VIII, 391 pages. 1985.

Vol. 19: Pictorial Information Systems in Medicine. Edited by K. H. Höhne. XII, 525 pages. 1986.

Vol. 20: Disordered Systems and Biological Organization. Edited by E. Bienenstock, F. Fogelman Soulié, G. Weisbuch. XXI, 405 pages.1986.

Vol. 21: Intelligent Decision Support in Process Environments. Edited by E. Hollnagel, G. Mancini, D.D. Woods. XV, 524 pages. 1986.

NATO ASI Series F

Vol. 22: Software System Design Methods. The Challenge of Advanced Computing Technology. Edited by J.K. Skwirzynski. XIII, 747 pages. 1986.

Vol. 23: Designing Computer-Based Learning Materials. Edited by H. Weinstock and A. Bork. IX, 285 pages. 1986.

Vol. 24: Database Machines. Modern Trends and Applications. Edited by A.K. Sood and A.H. Qureshi. VIII, 570 pages. 1986.

Vol. 25: Pyramidal Systems for Computer Vision. Edited by V. Cantoni and S. Levialdi. VIII, 392 pages. 1986. *(ROB)*

Vol. 26: Modelling and Analysis in Arms Control. Edited by R. Avenhaus, R.K. Huber and J.D. Kettelle. VIII, 488 pages. 1986.

Vol. 27: Computer Aided Optimal Design: Structural and Mechanical Systems. Edited by C.A. Mota Soares. XIII, 1029 pages. 1987.

Vol. 28: Distributed Operating Systems. Theory und Practice. Edited by Y. Paker, J.-P. Banatre and M. Bozyiğit. X, 379 pages. 1987.

Vol. 29: Languages for Sensor-Based Control in Robotics. Edited by U. Rembold and K. Hörmann. IX, 625 pages. 1987. *(ROB)*

Vol. 30: Pattern Recognition Theory and Applications. Edited by P.A. Devijver and J. Kittler. XI, 543 pages. 1987.

Vol. 31: Decision Support Systems: Theory and Application. Edited by C.W. Holsapple and A.B. Whinston. X, 500 pages. 1987.

Vol. 32: Information Systems: Failure Analysis. Edited by J.A. Wise and A. Debons. XV, 338 pages. 1987.

Vol. 33: Machine Intelligence and Knowledge Engineering for Robotic Applications. Edited by A.K.C. Wong and A. Pugh. XIV, 486 pages. 1987. *(ROB)*

Vol. 34: Modelling, Robustness and Sensitivity Reduction in Control Systems. Edited by R.F. Curtain. IX, 492 pages. 1987.

Vol. 35: Expert Judgment and Expert Systems. Edited by J.L. Mumpower, L.D. Phillips, O. Renn and V.R.R. Uppuluri. VIII, 361 pages. 1987.

Vol. 36: Logic of Programming and Calculi of Discrete Design. Edited by M. Broy. VII, 415 pages. 1987.

Vol. 37: Dynamics of Infinite Dimensional Systems. Edited by S.-N. Chow and J.K. Hale. IX, 514 pages. 1987.

Vol. 38: Flow Control of Congested Networks. Edited by A.R. Odoni, L. Bianco and G. Szegö. XII, 355 pages. 1987.

Vol. 39: Mathematics and Computer Science in Medical Imaging. Edited by M.A. Viergever and A. Todd-Pokropek. VIII, 546 pages. 1988.

Vol. 40: Theoretical Foundations of Computer Graphics and CAD. Edited by R.A. Earnshaw. XX, 1246 pages. 1988.

Vol. 41: Neural Computers. Edited by R. Eckmiller and Ch. v. d. Malsburg. XIII, 566 pages. 1988.

NATO ASI Series F

Vol. 42: Real-Time Object Measurement and Classification. Edited by A. K. Jain. VIII, 407 pages 1988. *(ROB)*

Vol. 43: Sensors and Sensory Systems for Advanced Robots. Edited by P. Dario. XI, 597 pages. 1988. *(ROB)*

Vol. 44: Signal Processing and Pattern Recognition in Nondestructive Evaluation of Materials. Edited by C.H. Chen. VIII, 344 pages. 1988. *(ROB)*

Vol. 45: Syntactic and Structural Pattern Recognition. Edited by G. Ferraté, T. Pavlidis, A. Sanfeliu, H. Bunke. XVI, 467 pages. 1988. *(ROB)*

Vol. 46: Recent Advances in Speech Understanding and Dialog Systems. Edited by H. Niemann, M. Lang, G. Sagerer. X, 521 pages. 1988.

Vol. 47: Advanced Computing Concepts and Techniques in Control Engineering. Edited by M.J. Denham, A.J. Laub. XI, 518 pages. 1988.

Vol. 48: Mathematical Models for Decision Support. Edited by Gautam Mitra. IX, 762 pages. 1988.

Vol. 49: Computer Integrated Manufacturing. Edited by I. Burhan Turksen. VIII, 568 pages. 1988.

Vol. 50: CAD Based Programming for Sensory Robots. Edited by B. Ravani. IX, 565 pages. 1988. *(ROB)*

Vol. 51: Algorithms and Model Formulations in Mathematical Programming. Edited by Stein W. Wallace. IX, 190 pages. 1989.

Vol. 52: Sensor Devices and Systems for Robotics. Edited by Alícia Casals. IX, 362 pages. 1989. *(ROB)*

Vol. 53: Advanced Information Technologies for Industrial Material Flow Systems. Edited by S.Y. Nof and C.L. Moodie. IX, 710 pages. 1989.

Vol. 54: A Reappraisal of the Efficiency of Financial Markets. Edited by R.M.C. Guimarães, B.G. Kingsman and S.J. Taylor. X, 804 pages. 1989.

Vol. 55: Constructive Methods in Computing Science. Edited by Manfred Broy. VII, 478 pages. 1989.

Vol. 56: Multiple Criteria Decision Making and Risk Analysis Using Microcomputers. Edited by B. Karpak and S. Zionts. VII, 399 pages. 1989.

MIX
Papier aus verantwortungsvollen Quellen
Paper from responsible sources
FSC® C105338

If you have any concerns about our products,
you can contact us on
ProductSafety@springernature.com

In case Publisher is established outside the EU,
the EU authorized representative is:
Springer Nature Customer Service Center GmbH
Europaplatz 3, 69115 Heidelberg, Germany

Printed by Libri Plureos GmbH
in Hamburg, Germany